The Art of Failure

Playful Thinking

Jesper Juul, Geoffrey Long, and William Uricchio, editors

The Art of Failure: An Essay on the Pain of Playing Video Games, Jesper Juul, 2013

The Art of Failure: An Essay on the Pain of Playing Video Games

Jesper Juul

The MIT Press
Cambridge, Massachusetts
London, England

MIT Press books may be purchased at special quantity discounts for business or sales promotional use. For information, please email special_sales@mitpress.mit.edu.

This book was set in Stone Sans and Stone Serif by Toppan Best-set Premedia Limited, Hong Kong. Printed and bound in the United States of America.

Library of Congress Cataloging-in-Publication Data

Juul, Jesper, 1970–
The art of failure : an essay on the pain of playing video games / Jesper Juul.
    p.   m.—(Playful thinking)
Includes bibliographical references (p.   ) and index.
ISBN 978-0-262-01905-7 (hardcover : alk.paper)
1. Video games—Psychological aspects. 2. Video games—Philosophy.
3. Failure (Psychology) I. Title.
GV1469.3.J87 2013
794.8—dc23
2012026836

10 9 8 7 6 5 4 3 2

To Otto, for teaching me how to play.

Contents

Series Foreword

Many people (we series editors included) find video games exhilarating, but it can be just as interesting to ponder *why* that is so. What do video games do? What can they be used for? How do they work? How do they relate to the rest of the world? Why is play both so important and so powerful?

Playful Thinking is a series of short, readable, and argumentative books that share some playfulness and excitement with the games that they are about. Each book in the series is small enough to fit in a backpack or coat pocket, and combines depth with readability for any reader interested in playing more thoughtfully or thinking more playfully. This includes, but is by no means limited to, academics, game makers, and curious players.

So, we are casting our net wide. Each book in our series provides a blend of new insights and interesting arguments with overviews of knowledge from game studies and other areas. You will see this reflected not just in the range of titles in our series, but in the range of authors creating them. Our basic assumption is simple: video games are such a flourishing medium that any new perspective on them is likely to show us something unseen or forgotten, including those from such "unconventional" voices as artists, philosophers, or specialists in other industries or fields of study. These books will be bridge-builders, cross-pollinating both areas with new knowledge and new ways of thinking.

At its heart, this is what Playful Thinking is all about: new ways of thinking about games, and new ways of using games to think about the rest of the world.

Preface

To begin with a confession: I am a sore loser. Something in me that demands that I win, beat, or complete every game I try, and that part of me is outraged and tormented whenever I fail to do so. Still, I play video games though I know I will fail, at least part of the time. On a higher level, I think I *enjoy* playing video games, but why does this enjoyment contain at its core something that I most certainly do not enjoy?

This book could not have been written without the help of friends and colleagues who challenged my arguments, fixed my mistakes, and offered me new ideas. My gratitude goes to those who commented on the manuscript as it evolved: Susana Tosca, Miguel Sicart, Markus Montola, Katherine Isbister, Ben Abraham, Jonathan Frome, Marie-Laure Ryan, Clara Fernández-Vara, Jan-Noël Thon, Chaim Gingold, and Bennett Foddy. To those who provided comments and ideas for the arguments as they developed: Albert Dang, Kan Yang Li, Frank Lantz, Eric Zimmerman, Nick Fortugno, Jason Begy, T. L. Taylor, Jonas Heide Smith, Chris Bateman, Matthew Weise, McKenzie Wark, Svend Juul, and the MIT Press reviewers. Needless to say, any remaining failures of this book are entirely my responsibility.

Thanks to Rachel Morris for illustrations and Charles Pratt for a super screenshot. Thanks to Geoffrey Long and William Uricchio, coeditors of the Playful Thinking series, as well as Doug Sery and the rest of the MIT Press for supporting the project.

Danish Centre for Design Research

The Art of Failure was made possible by a grant from the Danish Centre for Design Research and written during a stay at the Danish Design School in Copenhagen.

I would like to thank Troels Degn Johansson and the rest of the Danish Design School for giving me the opportunity to write, and to Parsons the New School of Design, the Singapore-MIT GAMBIT Game Lab, and the New York University Game Center, where many of the ideas that went into the manuscript were hatched. Also thanks to the speakers and participants at the *Unfortunate Game Events* seminar in Copenhagen.

Thanks to Nanna Debois Buhl.

Parts of chapter 5 were cowritten with Albert Dang and Kan Yang Li, and published as "The Suicide Game," in *Proceedings of the 2007 DiGRA Conference*, Tokyo, Japan.

Chapters 1–3 contain components of the article "Fear of Failing? The Many Meanings of Difficulty in Video Games," in *The Video Game Theory Reader 2*, ed. Bernard Perron and Mark J. P. Wolf (New York: Routledge, 2008).

The book uses material from the presentation "Beyond Balancing: Using Five Elements of Failure Design to Enhance Player

Experiences," presented at the *Game Developers Conference*, San Francisco, 2009.

Elements of chapter 3 were originally published as "In Search of Lost Time," in *Proceedings of the Fifth International Conference on the Foundations of Digital Games—FDG '10*, Monterey, California, 2010.

Additional material can be found on the book's Web site: http://www.jesperjuul.net/artoffailure

1 Introduction: The Paradox of Failure

I was playing *Patapon*[1] shown in figure 1.1. Things were going well, but when I came to the desert, my tactics began to fail. I repeated the trusted △-△-□-○ sequence of button pushes, but my warriors continued to burn to death in the sun; I failed the level; I tried again. I could not glean from the game if my timing was off, if I was using the wrong sequence, or if something completely different was wrong. I put the game away; I returned to it; I put it away again. I did not feel too good about myself. I dislike failing, sometimes to the extent that I will refuse to play, but mostly I will return, submitting myself to series of unhappy failures, once again seeking out a feeling that I deeply dread.

It is with some trepidation that I admit to my failures in *Patapon*, but I can fortunately share a story that puts my skills in a better light. I had been looking forward to *Meteos*[2] (figure 1.2) for a long time, so I unwrapped it quickly and selected the main game mode.[3] In a feat of gamesmanship (I believe), I played the game to completion on my very first attempt without failing even once. Naturally, this made me very angry. I put the game away, not touching it again for more than a year. (I have not been able to repeat this first performance.)

Figure 1.1
Patapon (Japan Studios 2008)

I dislike failing in games, but I dislike *not* failing even more. There are numerous ways to explain this contradiction, and I will discuss many of them in this book. But let us first consider the strangeness of the situation: every day, hundreds of millions of people around the world play video games, and most of them will experience failure while playing. It is safe to say that humans have a fundamental desire to succeed and feel competent,[4] but game players have chosen to engage in an activity in which they are almost certain to fail and feel incompetent, at least some of the time. In fact, we know that players prefer games in which they fail. This is the *paradox of failure* in games. It can be stated like this:

1. We generally avoid failure.

2. We experience failure when playing games.

3. We seek out games, although we will experience something that we normally avoid.

This paradox of failure is parallel to the paradox of why we consume tragic theater, novels, or cinema even though they

Figure 1.2
Meteos (Q Entertainment 2005)

make us feel sadness, fear, or even disgust. If these at first do not sound like actual paradoxes, it is simply because we are so used to their existence that we sometimes forget that they are paradoxes at all. The shared conundrum is that we generally try to avoid the unpleasant emotions that we get from hearing about a sad event, or from failing at a task. Yet we actively seek out these emotions in stories, art, and games.

The paradox of *tragedy* is commonly explained with reference to Aristotle's term *catharsis*, arguing that we in our general lives experience unpleasant emotions, but that by experiencing pity and fear in a fictional tragedy, these emotions are eventually purged from us.[5] However, this does not ring true for games—when we experience a humiliating defeat, we really are filled with emotions of humiliation and inadequacy. Games do not purge these emotions from us—they produce the emotions in the first place.

The paradox is not simply that games or tragedies contain something unpleasant in them, but that we appear to *want* this unpleasantness to be there, even if we also seem to dislike it (unlike queues in theme parks, for example, which we would prefer didn't exist). Another explanation could be that while we dislike failing in our regular endeavors, games are an entirely different thing, a safe space in which failure is okay, neither painful nor the least unpleasant. The phrase "It's just a game" suggests that this would be the case. And we *do* often take what happens in a game to have a different meaning from what is outside a game. To prevent other people from achieving their goals is usually hostile behavior that may end friendships, but we regularly prevent other players from achieving their goals when playing friendly games. Games, in this view, are something different from the regular world, a frame in which failure is not the least distressing. Yet this is clearly not the whole truth:

we are often upset when we fail, we put in considerable effort to avoid failure while playing a game, and we will even show anger toward those who foiled our clever in-game plans. In other words, we often argue that in-game failure is something harmless and neutral, but we repeatedly fail to act accordingly.

The reader has probably already thought of other solutions to the paradox of failure. I will discuss many possible explanations, and while I will propose an answer to the problem, the journey itself is meant to offer a new explanation of what it is that games *do*.

Players tend to prefer games that are somewhat challenging, and for a moment it can sound as if this explains the paradox—players like to fail, but not too much. Game developers similarly talk about *balancing*, saying that a game should be "neither too easy nor too hard," and it is often said that such a balance will put players in the attractive psychological state of *flow*[6] in which they become agreeably absorbed by a game.[7] Unfortunately, these observations do not actually explain the paradox of failure—they simply demonstrate that players and developers alike are aware of its existence. I will be discussing the paradox mostly in relation to video games (on consoles, computers, handheld devices, and so on), but it applies to all game types, digital or analog. I will also be looking at single-player games (failure against the challenge of the game), as well as competitive multiplayer games (failure against other players).

During the last few years, failure has become a contested discussion point in video game culture. Since roughly 2006, we have seen an explosion of new video game forms, with video games now being distributed not only in boxes sold in stores, but also on mobile phones, as downloads, in browsers, and on social networks, as well as being targeted at almost the entire population, and designed for all kinds of contexts for which

video games used to not be made. This *casual revolution*[8] in video
games is forcing us to rethink the role of failure in games: should
all games be intense personal struggles that bombard the player
with constant failures and frequent setbacks, or can games be
more relaxed experiences, like a walk in the park? The somewhat
anticipated response from part of the traditional video gaming
community has been to denounce new casual and social games
as too easy, pandering, simplistic, and so on. Yet, what has become clear is that both (a) many of the apparently simple games
played by a broad audience are in actuality very challenging and
(b) some traditional video game genres, especially role-playing
games, all but guarantee players that they will eventually pre-
vail. So failure is in need of a more detailed account, and we
must begin by asking the simple question: what *does* failure do?

Consider what happens when we are stuck in the puzzle
game *Portal 2*[9] shown in figure 1.3: we understand that we are

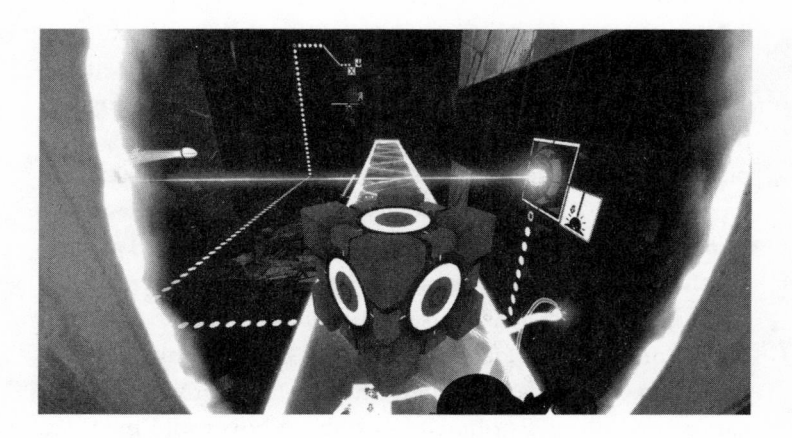

Figure 1.3
Portal 2 (Valve 2011)

lacking and inadequate (and more lacking and inadequate the longer we are stuck), but the game implicitly promises us that we can remedy the problem if we keep playing. Before playing a game in the *Portal* series, we probably did not consider the possibility that we would have problems solving the warp-based spatial puzzles that the game is based on—we had never seen such puzzles before! This is what games do: they promise us that we can repair a personal inadequacy—an inadequacy that they produce in us in the first place.

My argument is that the paradox of failure is unique in that when you fail in a game, it really means that *you* were in some way inadequate. Such a feeling of inadequacy is unpleasant for us, and it is odd that we choose to subject ourselves to it. However, while games uniquely induce such feelings of being inadequate, they also motivate us to play *more* in order to escape the same inadequacy, and the feeling of escaping failure (often by improving our skills) is central to the enjoyment of games. Games promise us a fair chance of redeeming ourselves. This distinguishes game failure from failure in our regular lives: (good) games are designed such that they give us a fair chance, whereas the regular world makes no such promises.

Games are also special in that the conventions around game playing are by themselves philosophies of the meaning of failure. The ideals of sportsmanship[10] specifically tell us to take success and failure seriously but to keep our emotions in check for the benefit of greater causes. Sports philosopher Peter Arnold has identified three types of sportsmanship: (1) sportsmanship as a form of social union (the noble behavior in the game extending outside the game), (2) sportsmanship as a means in the promotion of pleasure (controlling our behavior to make this and future games possible), and (3) sportsmanship as altruism (players

forfeiting a chance to win in order to protect another participant, for example).[11]

This type of emotional control can be challenging for children (and others), and a good deal of material exists for explain-
ing it. The book *Liam Wins the Game, Sometimes*[12] teaches children
how to deal with winning and losing in games. The author tells
the child that it is acceptable to feel disappointed when losing,
but unacceptable to throw a tantrum as shown in figure 1.4. "It
is being a poor loser and it spoils the whole game. Others do not

Figure 1.4
Teaching children how not to deal with failure. Illustration from Whelen
Banks, *Liam Wins the Game, Sometimes: A Story about Losing with Grace
(Liam Says)* (London and Philadelphia: Jessica Kingsley Publishers,
2009). Reproduced with permission of Jessica Kingsley Publishers.

like playing with poor losers."[13] To be a sore loser is to make a concrete philosophical claim: that failure in games is straightforwardly painful, without anything to compensate for it. However, it is important to realize that poor losers are not chastised for showing anger and frustration, but for showing anger and frustration in *the wrong way*. Games, depending on how we play them, give us a license to display anger and frustration on a level that we would not otherwise dare express, but some displays will still be out of bounds, rude, or socially awkward. Contrary to the poor loser, the spoilsport who plays a game without caring for either winning or losing is making the statement that game failure is not painful at all.

The Uses of Failure: Learning and Saving the World

Though we may dislike failure as such, failure is an integral element of the overall experience of playing a game, a motivator, something that helps us reconsider our strategies and see the strategic depth in a game, a clear proof that we have improved when we finally overcome it. Failure brings about something positive, but it is always potentially painful or at least unpleasant. This is the double nature of games, their quality as "pleasure spiked with pain."

This is why the question of failure is so important: it not only goes to the heart of why we enjoy games in the first place, it also tells us what games can be used for. Given that games have an undisputable ability to motivate players to meet challenges and learn in order to overcome failure, wouldn't it be smart to use games to motivate players toward other more "serious" undertakings?[14] It is commonly argued that the principles of game design can be applied to a number of situations in the regular world in order to motivate us: examples include designing

educational games, giving employees points for their performance, giving shoppers points for checking in at specific loca-
tion, awarding Internet users with badges for commenting on
Web site posts, and so on. This is a long-standing idea, which
at the time of writing has resurfaced under the name of *gamifica-
tion.*[15] We therefore need to think more closely about why games
work so well: at the very least, good games tend to offer well-
defined goals and clear feedback. This gives us an objective
measure of our performance, and allows us to optimize our
strategies. If applying this to nongame situations sounds tempting, consider how the 2008 financial crisis was caused in part
by large banks and financial institutions making their organizations too gamelike by giving employees the clear goal of approv-
ing as many loans as possible and punishing naysayers with
termination. This was a case where the design that works so well
inside games can be disastrous outside games, even if we think
only of the well-being of the companies involved. Games, apparently, are not a pixie dust of motivation to be sprinkled on any
subject. The underlying questions are therefore: When and how
do games motivate us to overcome failure and improve ourselves? When is a game structure useful, and when is it detri-
mental? And most important: Is there a difference between
failing inside and failing outside a game?

Inside and Outside the Game

Imagine that you are dining with some people you have just
met. You reach for the saltshaker, but suddenly one of the other
guests, let's call him Joe, looks at you sullenly, then snatches
the salt away and puts it out of your reach. Later, when you are
leaving the restaurant, Joe dashes ahead of you and blocks the

exit door from the outside. Joe is being rude—when you understand what another person is trying to do, it is offensive, or at least confrontational, to prevent that person from doing it.

However, if you were meeting the same people to play the board game *Settlers*,[16] it would be completely acceptable for the same Joe to prevent you from winning the game. In the restaurant as well as in the game, Joe is aware of your intention, and Joe prevents you from doing what you are trying to do. At the restaurant, this is rude. In the game, this is expected and acceptable behavior. Apparently, games give us a license to engage in conflicts, to prevent others from achieving their goals. When playing a game, a number of actions that would regularly be awkward and rude are recast as pleasant and sociable (as long as we are not poor losers, of course).

Similarly, consider how the designer of a car, computer program, or household appliance is obliged to make sure that users find the design easy to use. At the very least, the designer is expected to help the driver avoid oncoming traffic, prevent the user from deleting important files, and not trick the user into selecting the wrong temperature for a wash.[17] A fictional example shows what can happen if designers do not live up to his obligation: in Monty Python's "Dirty Hungarian Phrasebook" sketch, a malicious author creates a fake Hungarian language phrasebook in which (among other things) a request for the way to the train station is translated into English as sexual innuendo.[18] Chaos ensues. We expect neither phrasebook authors nor designers to act this way.

However, if you pick up a single-player video game, you expect the designer to have spent considerable effort preventing you from easily reaching your goal, all but guaranteeing that you will at least temporarily fail. (Designers are also expected

Figure 1.5
One-button game

to make *some* parts of a game easy to use.[19]) It would be much
easier for the designer to create the game shown in figure 1.5,
where the user only has to press a button once to complete the
game. But for something to be a good game, and a game at all,
we expect resistance and the possibility of failure. Single- and
multiplayer games share this inversion of regular courtesy, giving
players license to work against each other where it would otherwise be rude, and allowing the designer to make life difficult for
the player.

If we return to Joe, the rude dinner companion who denied
you access to the salt and blocked the door, we could also
imagine him performing the very same actions with a glimmer
in his eye, smiling, and perhaps tilting his head slightly to the
side. In this case, Joe is not trying to be rude, but *playful*, and
you may or may not be willing to play along. By performing
simple actions such as saying "Let's play a game" or tilting our
heads and smiling, we can change the expectations for what is
to come. Gregory Bateson calls this *meta-communication*: humans

and other animals (especially mammals) perform playful actions where, for example, what looks like a bite is understood to not be an actual bite.[20] Such meta-communication is found in all types of play, but games are a unique type of structured play that allows us to perform seemingly aggressive actions within a frame where they are understood as not quite aggressive.

In the field of game studies, Katie Salen and Eric Zimmerman have described game playing as entering a *magic circle* in which special rules apply.[21] This idea of a separate space of game playing has been criticized on the grounds that there is no perfect separation between what happens inside a game, and what happens outside a game.[22] That is obviously true but misses the point: the circumstances of your game playing, personality, mood, and time investment will influence how you feel about failure, but we nevertheless treat games differently from non-games, and we have ways of initiating play. We expect certain behaviors and experiences within games, but there are no guarantees that players, ourselves included, will live up to expectations.

The Gamble of Failure

It's easy to tell what games my husband enjoys the most. If he screams 'I hate it. I hate it. I hate it,' then I know he will finish it and buy version two. If he doesn't say this, he'll put it down in an hour.[23]

In quoting the spouse of a video game player, game emotion theorist Nicole Lazzaro shows how we can be angry and frustrated while playing a game, but that this frustration and anger binds us to the game. We are motivated to play when something is at stake. It seems that the more time we invest into overcoming a challenge (be it completing a game, or simply overcoming a small subtask), the bigger the sense of loss we experience when failing, and the bigger the sense of triumph we feel when succeeding. Even then, our feeling of triumph can quickly evaporate if we learn that other players overcame the challenge faster than we did. To play a game is to make an emotional gamble: we invest time and self-esteem in the hope that it will pay off. Players are not willing to run the same amount of risk—some even prefer not to run a risk at all, not to play.

I am taking a broad view of failure here. Examples of failures include the GAME OVER screen of a traditional arcade game such as *Pac-Man*[24] (figure 1.6), the failure of a player to complete a level within sixty seconds, the failure to survive an onslaught of opponents, the failure to complete a mission in *Red Dead Redemption*,[25] the failure to protect the player character in *Limbo*,[26] the failure to win a tic-tac-toe match against a sibling, and the failure to win Wimbledon or the Tour de France. It can also be something as ordinary as the failure to jump to the next ledge in a platform game like *Super Mario Bros.*,[27] even when it has no consequences beyond having to try the jump again (figure 1.7). Though on different scales, each of these examples involves *the player working toward a goal, either communicated by the game or invented by the player, and the player failing to attain that goal*. Depending on the goal of a given game, failures can result in either a permanent loss (such as when losing a match in multiplayer game) or a loss of time invested toward completing or progressing in a game.

Certainly, the experience of failing in a game is quite different from the experience of witnessing a protagonist failing in a story. When reading a detective story, we follow the thoughts and discoveries of the detective, and when all is revealed, nothing prevents us from believing that we had it figured out all along. Through fiction, we can feel that we are smart and successful, and stories politely refrain from challenging that belief. Games call our bluff and let us know that we failed. Where novels and movies concern the personal limitations and self-doubt of others, games have to do with *our* actual limitation and self-doubts. However much we would like to hide it, our failures are plain to see for any onlooker, and any frustration that we indicate is easily understood by anyone who watches us.

But, we often identity

This Game Is Stupid Anyway *w/ the characters in the books*

"This Sport Is Stupid Anyway," a New York Post headline proclaimed following the US soccer team's exit from the 2010 World Cup.[28] Fortunately, we have ways of denying that we care about failure. We can dismiss a game as poorly made or even "stupid," and we understand this type of defense to be so childish that we will use it only half-jokingly as in the *New York Post* headline. This is an opportunistic "theory" about the paradox of failure: that failure in a specific game is unimportant, because it requires only irrelevant skills (if any).

Having failed in *Patapon*, I searched for "Patapon desert" and learned that I needed a "JuJu," a rain miracle which I did not recall having ever heard of. To my great relief, the search yielded more than 150,000 hits—I was not the only player to suffer from this problem, and I could safely conclude that the problem lay with the game, certainly not with me. Our experience of failure strongly depends on how we assign the blame for failing. In psychology, *attribution theory* explains that we try to attribute events to certain causes. Harold K. Kelley distinguishes among three types of attributions that we can make in an event involving a person and an entity.[29]

attribution theory

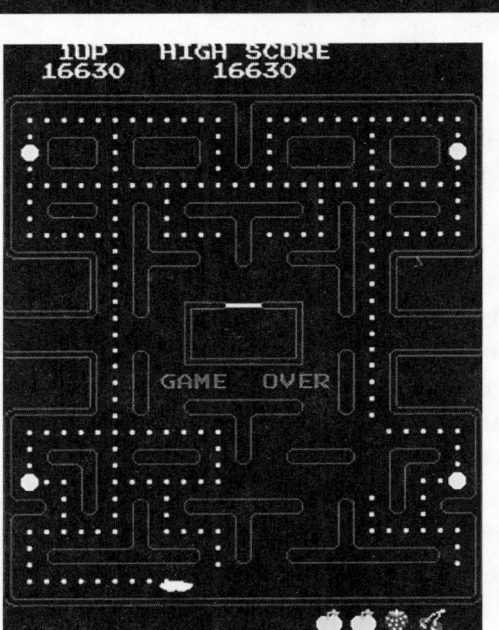

Figure 1.6
Failures: *Limbo* failed puzzle (Playdead Studios 2010); *Pac-Man* GAME
OVER (Namco 1980); *Super Monkey Ball Deluxe* fallout (Sega 2005)

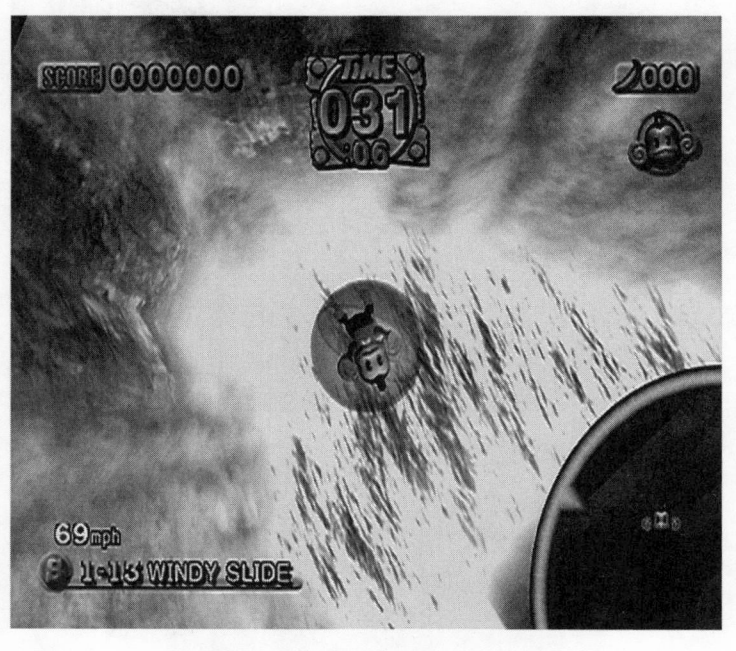

Figure 1.6
(continued)

• *Person:* The event was caused by personal traits, such as skill and disposition.

• *Entity:* The event was caused by characteristics of the entity.

• *Circumstances:* The event was due to transient causes such as luck, chance, or an extraordinary effort from the person.

If we receive a low grade on a school test, we can decide that this was due to (1) person—personal disposition such as lack of skill, (2) entity—an unfair test, or (3) circumstance—having slept badly, having not studied enough. This maps well to common explanations for failure in video gaming: a player who loses a

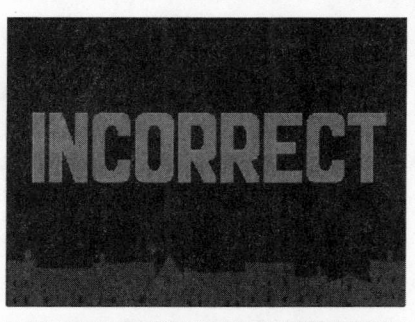

Figure 1.7
Lost StarCraft II match (Blizzard Entertainment 2010); *Professor Layton* failed puzzle (Level-5 2008); *Super Mario Bros.* failed jump (Nintendo 1985)

Figure 1.7
(continued)

game can claim to be bad at this specific game or at video games in general, claim that the game is unfair, or dismiss failure as a temporary state soon to be remedied though better luck or preparation.

I blamed *Patapon*: I searched for a solution, and I used the fact that many players had experienced the same problem as an argument for attributing my failure to a flaw in the game design, rather than a flaw with my skills. As it happens, we are a self-serving species, more likely to deny responsibility when we fail than when we succeed. A technical term for this is *motivational bias*, but it is also captured in the observation that "success has many fathers, but failure is an orphan." After numerous attempts

at this section of *Patapon*, I was relieved to be allowed to be
furious *at the game*, which I could now declare to be so poorly
designed that it was not worth my time. I put the game back in
its box, only returning to it months later. While we dislike
feeling responsible for failure, we dislike even more strongly
games in which we do not feel responsible for failure (a variation
on the fact that we do not want to fail in a game, but we also
do not want not to fail). The times I denied responsibility for
failure in *Patapon* and stopped playing, I precluded the possibility that I would eventually cross the desert and complete the
game. By refusing the emotional gamble of the game, I was
acting in a *self-defeating* way; by refusing to exert effort in order
to progress in the game, I was shielding myself from possible
future failures. According to one theory, our fear of failure leads
to procrastination: we perform worse than we should in order
to feel better about our poor performance.

Still, should we accept responsibility for failure, the question
becomes this: does my in-game performance reflect skills or
traits that I generally value? Benjamin Franklin notably declared
chess to be a game that contains important lessons: "The game
of Chess is not merely an idle amusement. Several very valuable
qualities of the mind, useful in the course of human life, are to
be acquired or strengthened by it, so as to become habits, ready
on all occasions . . . we learn by Chess the habit of not being
discouraged by present appearances in the state of our affairs,
the habit of hoping for a favourable change, and that of perse-
vering in the search of resources."[30] If we praise a game for
teaching important skills, as Franklin does here, we must accept
that failing in it will imply a personal lack of the same important skills. That is a question to ask about every game: does this
game expose our important underlying inadequacies, or does

it merely create artificial and irrelevant ones? If a game exposes existing inadequacies, then we must fear how it reveals our hidden flaws. If, rather, a game creates new, artificial "art" inadequacies, it is easier to shrug off.

Every failure we experience in a game is torn between these two arguments pulling in opposite directions: we can think of game failure as *normal*: as a type of failure that genuinely reflects our general abilities and therefore is as important as any out-of-game failure. However, we can also think of it as *deflated*: that the importance of any failure is automatically deflated when it occurs inside a game, since games are artificial constructs with no bearing on the regular world. My point is not that these two arguments are true or false, as much as games work by making these contradictory views available to us: failure really does matter to us, as can be witnessed in the way we try to avoid failure while playing and in the way we sometimes react when we do fail. At the same time, we use deflationary arguments to protect our self-esteem when we fail, and this gives games a kind of lightness and freedom that allows us to perform to the best of ability, because we have the option of denying that game failure matters.

The Meaning of the Art Form

Even if we often dismiss the importance of games, we also discuss them, especially the games that we call sports, as something above, something more pure than, everyday life. In professional sports, games are often framed as something noble, something that truly reveals the best side of humans, something larger than life—think only of movies like *Chariots of Fire*,[31] or the cultural obsession with athletes. In soccer, the Real Madrid–Barcelona rivalry continues to be played out with a layer of meaning that goes back to the Franco era. In baseball, the New York Yankees and the Boston Red Sox have competed for over a century, and every match between the two teams is seen through that lens and adds to that history. This extends beyond games involving physical effort. For example, the legendary 1972 World Chess Championship match in Reykjavik between US player Bobby Fischer and Soviet Boris Spassky was understood as an extension of the Cold War. These examples demonstrate that we routinely understand games as more important, more glorious, and more tragic than everyday life.

Outside the realm of sports, late eighteenth-century German philosopher Friedrich Schiller went so far as to declare *play* central to being human. "Man plays only when he is in the full sense of the word a man, and *he is only wholly Man when he is playing.*"[32] In the 1930s, Dutch play theorist Johan Huizinga noted this duality between our framing of games as either important or frivolous, by describing play as "a free activity standing quite consciously outside 'ordinary' life as being 'not serious', but at the same time absorbing the player intensely and utterly."[33] We can talk about games as either carved-off experiences with no bearing on the rest of the world or as revealing something deeper, something truly human, something otherwise invisible.

This type of discussion, of whether game failures, and games by extension, are significant, has been applied to every art form. All humans consume artistic expressions from music through storytelling to the visual arts. We may share the intuition that the arts are fruitful, inspirational, and important, yet it is hard to demonstrate or measure such positive effects. In *The Republic*, Plato famously denied the poet access to his ideal society "because he wakens and encourages [Plato] and strengthens the lower elements in the mind to the detriment of reason."[34] Compare this with the continued idealization of art as a privileged way of understanding the world.[35] Games share this predicament with other art forms: we may sense that they are important, that they give access to something profound; it is just that we have no easy way to prove that. Games are activities that have no *necessary* tangible consequences (though we can negotiate to play for concrete consequences—money, doing the dishes, etc.).[36] This lack of necessary tangible consequences (productive, negative, or positive) defines games, but it can also make them seem frivolous.[37] Yet it is precisely because games are not obviously necessary for our daily lives that we can declare them to be above the banality of our simpler, more mundane needs.

Video games have by now celebrated their fiftieth anniversary, while games in general have been around for at least five thousand years. The first decade of this century saw the appearance of the new field of video game studies, including conferences, journals, and university programs. The defense of video games (as of most things) tends to grow from personal fascination. *I* enjoy video games; *I* feel that they give me important experiences; *I* associate them with wide-ranging thoughts about life, the universe, and so on. This is valuable *to me*, and *I* want to understand and share it. From that starting point, video game fans have so far focused on two different arguments for the value of video games:

1. *Video games can do what established art forms do.* In this strategy, the fan claims that video games can produce the same type of experiences as (typically) cinema or literature produce. Are video games not engaging like *War and Peace* or *The Seventh Seal*? The downside to this strategy is that it makes video games

sound derivative if we only argue that video games live up to criteria set by literature or cinema, why bother with games at all?

2. *Video games transcend established categories.* In this strategy, the fan can argue that since we already have film, why should video games aspire to be film? It follows that we need to identify and appraise the *unique* qualities of video games. In its most austere form, this can become an argument for identifying a "pure" game that should be purged of influences from other art forms, typically by banishing straightforward narrative from game design. The softer version of this argument (which happens to be my personal position) states that video games should try to explore their own unique qualities, while borrowing liberally from other art forms as needed.

Again, these are theories that we use to explain our experiences. When I play video games, I do experience something important, profound. Video games are for me a space of reflection, a constant measuring of my abilities, a mirror in which I can see my everyday behavior reflected, amplified, distorted, and revealed, a place where I deal with failure and learn how to rise to a challenge. Which is to say that video games give me unique and valuable experiences, regardless of how I would like to argue for their worth as an art form, as a form of expression, and so on. With this book, I hope to bring the experience and the arguments closer to each other.

Two Types of Failure (and Tragedy)

In my earlier book *Half-Real*,[38] I argued that nonabstract video games are two quite different things at the same time: they are real rule systems that we interact with, and they are fictional worlds that the game cues us into imagining. For example, to win or lose a video game is an actual, real event determined by the game rules, but if we succeed by slaying a night elf, that adversary is clearly imaginary. As players, we switch between these two perspectives, understanding that some game events are part of the fictional world of the game (Mario's girlfriend has been kidnapped), while other game events belong to the rules of the game (Mario comes back from the dead after being hit by a barrel). This also means that there are two types of failure in games: *real failure* occurs when a player invests time into playing a game and fails; *fictional failure* is what befalls the character(s) in the fictional game world.

Real Failure
Like tragedy in theater, cinema, and literature, failure makes us experience emotions that we generally find unpleasant. The difference is that games can be tragic in a literal sense: consider the case of French bicycle racer Raymond Poulidor, who between 1962 and 1976 achieved no less than three second places and five third places in the Tour de France, but in his career never managed to win the race. Tragic.

On the other hand, if I fail to complete one level of a small puzzle game on my mobile phone because I have to get off at the right subway stop, we probably would not describe this as tragic. Not because there is any structural difference between the two situations—Poulidor and I both tried to win a game, and we both failed. We had both invested some time in playing, we had both made an emotional gamble in the hope that we would end up happy, and we both experienced a sense of loss when failing. Yet it is safe to say that Poulidor made a larger time investment and a larger emotional gamble than I did.

Playwright Oscar Mandel's traditional but often-cited definition of tragedy explains the difference between Poulidor and me: "A work of art is tragic if it substantiates the following situation: a protagonist who commands our earnest goodwill is impelled in a given world by a purpose, or *undertakes some action, of a certain seriousness and magnitude*; and by that very purpose or action, subject to the same given world, necessarily and inevitably meets with great spiritual or physical suffering" (my emphasis).[39] We reserve the idea of tragedy for events of some magnitude: my failing at a simple puzzle game does not qualify as tragic, but Poulidor's failed lifetime project of winning the Tour de France does. ✗ tragedy

Games are meaningful not simply by representing tragedies, but on occasion by creating actual, personal tragedies. In *The Birth of Tragedy*, Nietzsche discusses the notion that tragedy adds a layer of meaning to human suffering, that art "did not simply imitate the reality of nature but rather supplied a metaphysical supplement to the reality of nature, and was set alongside the latter as a way of overcoming it."[40] Though I am of a more optimistic temperament than Nietzsche was, I believe that there is a fundamental truth to this idea. Not in the naïve romantic sense that tragic themes are required for art to be valuable, but in the sense that painful emotions in art (such as games) gives us a space for contemplating the very same emotions. To some it may be surprising to hear that video games provide a space for contemplation at all, but it is probably more obvious when we consider that video games are part of an at least five-thousand-year history of games. Games, in turn, are often ritualistic, repeatable, and laden with symbolic meaning. Think only of Chess, or Go, or the Olympics. Or, casting an even wider net, play theorist Brian Sutton-Smith has proposed that play is fundamentally a "parody of emotional vulnerability": that through play we experience precarious emotions such as anger, fear, shock, disgust, and loneliness in transformed, masked, or hidden form.[41]

Fictional Failure

That was the real, first-person aspect of failure. We are real-life people who try to master a game, but most video games represent a mirror of our performance in their fictional worlds—they ask us to make things right in the game world by saving someone or fighting for self-preservation. For example, the game *Mass Effect 2*[42] lets the player steer Commander Shepard through a series of missions, protecting Shepard from harm and attempting to save the galaxy. The goals of the player are thus aligned with the goals of the protagonist; when the player succeeds, the protagonist succeeds. In games with no single protagonist, the player is typically asked to guard the interests of a group of people, a city, or a world.

The question is, can we imagine video games where this is inverted, such that when the player is successful, the protagonist fails? In the early 2000s, this seemed obviously impossible. As fiction theorist Marie-Laure Ryan put it, who would want to play *Anna Karenina*, the video game? Who would want to spend hours playing in order to successfully throw the protagonist under a train?[43] At the time, I also believed that such a game was inconceivable.[44] But only a few years later, there were games in which players had to do exactly that—kill themselves. Some of these were parodic games that openly subverted player expectations. Others were tragic in a traditional sense (SPOILER ALERT): *Red Dead Redemption*,[45] shown in figure 1.8, at first seems to let the player be a common video game hero, but the game can in

Figure 1.8
Red Dead Redemption (Rockstar Games 2010)

fact be completed only by sacrificing the protagonist in order
to save his family.

Director Steven Spielberg has argued that video games will
only become a proper *storytelling* art form "when somebody
confesses that they cried at level 17."[46] This is surely too simple:
any checklist for what makes a work of art *good* will necessarily
miss its mark, and works created to tick off the boxes in such a
list are rarely worthy of our attention. Ironically, the fact is that
players often do cry over video games, but mostly over losing
important matches in multiplayer games, being expelled from
their guild in *World of* Warcraft,[47] and so forth. Players report
crying over some single-player games such as *Final Fantasy VII.*[48]
Note that tragic endings in games are not interesting because they
magically transform video games into a respected art form, but
because they show that games can deal with types of content
that we thought could not be represented in this form. Tragic

game endings appear distressing due to the tension between the success of the player and the failure of a game protagonist, but this distress can give us a sense of responsibility and complicity, creating an entirely new type of tragedy.

Paths to Success

Games are not just about failure, of course. The general contract of game playing is that we promise we will be unhappy if we fail, and happy if we succeed in a game (even if we are playing a game that cannot be completed). We generally ask that games be *fair* in our quest for success, but, as it turns out, our idea of fairness has changed dramatically in the history of video games: whereas early arcade games could not actually be completed, home computer and console games gradually gained reachable endings during the 1980s and 1990s. By now, commercial single-player video games tend to come with the promise that almost all players will be able to complete them, though for different reasons in different games. As I write this, there is a common perception that video games are becoming *too* easy. Part of the issue is the rise of a type of game design that rewards mere time investment rather than skill. This type of design is most commonly criticized in social network games such as *FarmVille*,[19] but the truth is that the principle of directly rewarding the player for time invested is also common in role-playing games, digital or analog. A vocal group of players and developers are revolting against this, describing it as a watering down of video games, demanding that games should be "fair" by rewarding skill development, not time investment. In short, these dissenters think that games should make us fail more, but that success and failure should exclusively be a function of player skills.

In This Book

Failure in games tells us that we are flawed and deficient. As such, video games are the *art of failure*, the singular art form that sets us up for failure and allows us to experience and experiment with failure. I am going to make my argument through the four different lenses of philosophy, psychology, game design, and fiction, and these will be combined with my own experiences as a game player, and with experimental video games created for examining failure. The reader may note that I am quite promiscuous in my choice of disciplines and methods, but I am convinced that this will take us much further than if I had limited my analysis to a single tradition.

Following this introductory first chapter, chapter 2 examines failure in games using philosophical explanations of the paradox of tragedy. I argue that games have the double character of being both immediately personal and something whose importance we (often falsely) have an opportunity to dismiss. I furthermore examine how game playing conventions are by themselves philosophies of the paradox of failure.

Chapter 3 examines the psychology of failure. The paradox here returns as the contradictory demands of caring sufficiently about succeeding to put in effort, while not taking failure so seriously as to be discouraged or lose focus. The chapter shows that failure is perceived as more negative if it is communicated harshly, or if the player believes that progress can never be achieved. Furthermore, we often play badly to avoid feeling responsible for failure.

Chapter 4 examines how games make us fail, and how they promise that we can escape failure by the paths of skill, chance, or labor (time commitment). Each path makes failure personal in its own way: failing by skill is a pure personal flaw; failing by chance is either inconsequential or of cosmic implications (no gods are smiling upon the player); failing by labor indicates a lack of commitment. In addition, games have three types of goals that we can fail against: Games with transient goals can never be completed, and failing is therefore only a failure to win *this* match. Games with completable goals can be completed once and for all, but also let us continually fail against the overall goal of completion. Improvement goals are about our personal and continuous quest for improvement, and failure is simply a temporary failure to improve.

Chapter 5 shows that games can also be tragic in their fictions: fictional tragedy in games is a counterintuitive combination of failure and success—the failure of the fictional protagonist(s), but the success of the real player. Are players willing to play for the goal of the self-destruction of the protagonist, or are games better suited for other types of tragic content? The chapter shows that the most effective way of dealing with tragic content in game form is by making the player feel complicit with unpleasant events.

Finally, chapter 6 sums up the paradox of failure and argues for the uniqueness of games as an art form concerned with failure. Game structures are powerful motivators and guides for player behavior, but they cannot in every case be applied to other areas of life (work, business). Games are unanchored activities, with no necessary tangible consequences, and a fundamental unclearness about what it means to fail. Once an activity becomes tied to clear consequences for failure, the strength of games is quickly lost. Finally, while games provide a space in which we can experiment with failure, we should always grant ourselves one important right: the right to be genuinely frustrated when failing.

2 The Paradox of Failure and the Paradox of Tragedy

I possess a profound love for playing games, but when I fail, the last thing I feel is love. Therein lies the paradox of failure:

1. We generally avoid failure.

2. We experience failure when playing games.

3. We seek out games, although we will likely experience something that we normally avoid.

This is a paradox because statements 1 and 2 both appear reasonable, yet they lead to a counterintuitive conclusion when combined in statement 3.

The paradox of failure could also be called "Why am I doing this?" When experiencing an unpleasant failure in a game, yet desperately wanting to continue playing, why do I want to experience something that I also do not want to experience? This paradox of why we submit ourselves to failure is not entirely identical to the paradox of why we submit ourselves to tragedies, but the differences give some hints as to how video games differ from more established forms of culture. The paradox of failure concerns the concrete activity of playing a game and having our performance measured by the game. Chapter 5 deals with the separate issue of how we deal with failure in the fictional world of a game.

Figure 2.1
Thomas Rowlandson, *A Hitt at Backgammon*. EB8 R7967 810h, Houghton Library, Harvard University.

An obvious way of resolving the paradox would be to claim that failure in games is never the least bit painful or unpleasant. If so, games are all bliss, and the paradox disappears. Figure 2.1 shows why this argument is flawed: Players often exhibit great frustration when failing. Furthermore, players clearly exert effort in order to avoid failure. (Chapter 3 discusses the creative and sometimes destructive exceptions to this rule.) We may say that we do not mind failure in a game, but our frustration when failing and our in-game behavior both suggest otherwise.

Nevertheless, even though players appear to dislike failure, we tend to believe that games should make players fail, at least some of the time. From my own game playing and game design observations, I share this belief, but is it true? In collaboration

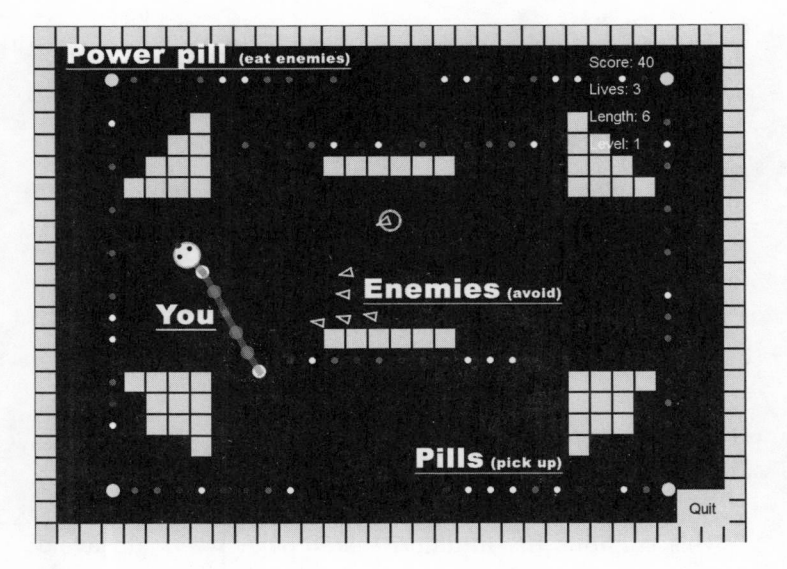

Figure 2.2
A game to test if failure is correlated with enjoyment

with the game company Gamelab, I developed a game that combines *Pac-Man*[1] and *Snake*: using the mouse, the player controls a snake that grows as the player collects pills; the player must avoid opponents; a special power pill allows the player to attack opponents for a short while (figure 2.2).

I recruited eighty-five players online[2] who played the game and were asked to rate it afterward. Figure 2.3 compares how well players did with how they rated the game. As it turned out, the most positive players were the ones who failed some, and then completed the game.[3] Players who completed the game without failing gave it a *lower* rating than those who failed at least once.[4] (As the ratings show, it was not a great game, but that is beside the point here.)

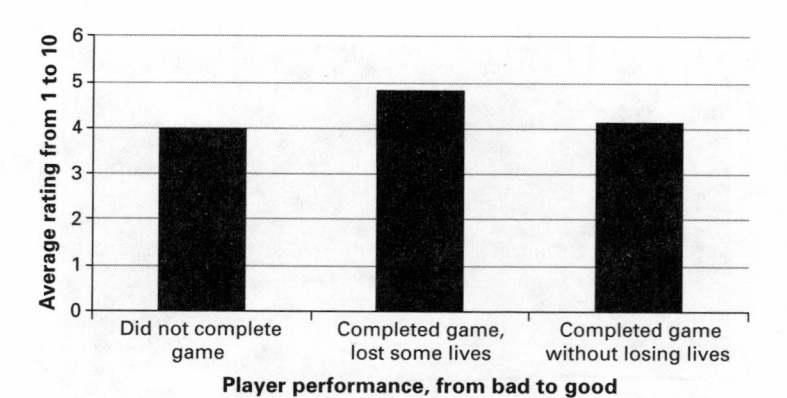

Figure 2.3
Players rate a game higher if they fail at least once

This confirms the intuition that though we try to avoid failure while playing, failure nevertheless gives a positive contribution to our evaluation of a game. There is something in games that we appear not to enjoy, but that coincides with us appreciating a game more. Unfortunately, this study also demonstrates only that the paradox exists, but does not solve it. Fortunately, we do not have to start from scratch but can learn from another discussion, one that has been going on for a few thousand years.

The Paradox of Painful Art

The paradox of failure and the better-known paradox of tragedy are part of a larger complex that has been called the *paradox of painful art*. Philosopher Aaron Smuts has described the general paradox as follows:

1. People do not seek out situations that arouse painful emotions.

2. People have painful emotions in response to some art.

3. People seek out art that they know will arouse painful emotions.[5]

This is a genuine paradox since some of our basic assumptions about human behavior (that humans avoid painful emotions) conflict with how humans actually behave (humans appear to seek out painful emotions in art). This means that something must be wrong with our assumptions or our reasoning, but what? As it turns out, the history of philosophy is littered with attempts to solve the paradox, and even providing an overview of such papers on the subject would vastly exceed the space I have here. In a survey of the paradox, Smuts divides the traditional solutions into three types:[6]

1. *Deflation: Art is not painful.* This type of solution agrees with the first premise of the paradox (that we avoid pain), but it denies the second premise (that we experience pain in relation to art).

2. *Compensation: Pain is compensated for.* This type of solution agrees with first two premises of the paradox (humans avoid pain and we experience genuine pain in relation to art), but proceeds to argue that art provides something positive that compensates for the pain. *We do not always seek*

3. *A-hedonism: We do not always seek pleasure.* This type of solution denies the first premise of the paradox by saying *pleasure.* that humans are not simply pain-avoiding, pleasure-seeking creatures. This is an effective, if unusual, way of dissolving the paradox.

I will begin by discussing the solutions in general, and then consider how they apply to failure in games.

1.   Deflation: Art is not painful.

The first type of solution is the *deflationary*[7] argument, which denies that art gives us painful emotions at all. According to Kendall Walton, we are only *fictionally* feeling sad when we experience a tragic story.[8] Of course, Walton's argument has some similarity to Samuel Taylor Coleridge's argument that in order to believe in poetry, we must exhibit a "willing suspension of disbelief."[9] In this line of reasoning, we intuitively disbelieve the worlds that artworks present to us, but we make an effort to disregard this disbelief in order to engage with fictions and to have an emotional response to them. This explanation is not quite satisfying: it seems odd to claim that we make a conscious effort to believe in a fiction when the common experience is rather that of being sucked in by a novel or a film, no will or effort required. Robert Yanal has argued that Walton and Coleridge are wrong to assume that belief is necessary for creating emotions (and there is therefore no need to suspend any disbelief). We simply have emotional responses to events regardless of whether we believe them to be true or not.[10]

As a variation on the deflationary argument, *control* theorists claim that because we have the option to turn off the television, stop reading the book, or leave the theater, we are protected from experiencing *genuinely* painful emotions. In this view, art is special because it grants us a kind of control that we lack outside art, and this control protects us from experiencing genuine pain.[11]

Another subspecies of deflation, *conversionary* arguments argue that our relation to an artwork changes over time because the work of art offers qualities (such as artistic qualities) that transform the initially negative emotions into something else. David Hume famously contends that the artistic qualities of a tragedy transform painful emotions into something positive: that we experience pleasure over the "very excellence with which the melancholy scene is represented."[12] This line of reasoning does not deny the existence of painful emotions in relation to art *in the first place*, but it claims that these emotions go away over time, such that we leave the artwork with no important experience of pain.

2. Compensation: Pain is compensated for.

Compensation is the second type of explanation. According to compensatory arguments, we experience genuinely unpleasant emotions in response to art, but this unpleasantness happens to be outweighed by positive factors.

The idea of *catharsis* is probably the most popular explanation of the paradox of painful art, but it is only seldom used in the realm of philosophy, and is furthermore quite enigmatic since Aristotle only mentions catharsis once in the *Poetics*: "Tragedy, then, is mimesis of an action which is elevated, complete, and of magnitude; in language embellished by distinct forms in its sections; employing the mode of enactment, not narrative; and through pity and fear accomplishing the catharsis of such emotions."[13] Writers on Aristotle have spent centuries in agitated argument about the true meaning of this passage, but for this purpose, the point is that catharsis has most commonly been understood as a type of *purgation*.[14] According to this, we may have experienced sadness in our lives, but by watching a fictional tragedy that sadness is purged from us. We experience something genuinely unpleasant but in turn receive something positive that compensates for this unpleasantness.

Film theorist Noël Carroll's theory of horror does not claim that we are purged of the feeling of horror, but that we receive something else instead. Carroll argues that when watching a horror movie, we genuinely dislike being horrified, but this unpleasantness is outweighed by the cognitive enjoyment of learning more about the enigmatic monster at the center of the story. The experience of horror is a simply a price we are willing to pay in order to reach the enjoyment of learning about the monster.[15] The Other

Other variations on the compensatory argument suggest that we feel good about feeling pity for someone else,[16] or about being able to endure unpleasant emotions.

3. A-hedonism: We do not always seek pleasure.

The third, less common, solution is a-hedonism, which denies the first premise of the paradox and argues that humans do not simply (or primarily) seek pleasure. For example, Alex Neill argues that "the idea that we commonly enjoy or take pleasure in seeing Oedipus or Gloucester stumbling around with their eyes out is after all somewhat peculiar,"[17] and that "pleasure" is an incorrect description of our reaction to tragedies. He believes that *satisfaction* is a better term since it does not appear paradoxical when describing a painful response.[18] Neill goes on to claim that we do experience pleasure from horror (think of people laughing, having watched a horror movie), but not from tragedy, the traditional subject of the paradox.

The Paradox of Failure and the Paradox of Painful Art

How does this apply to failure? Writing this small survey of solutions to the paradox gives me something close to an "all of the above" experience. Not every variation of every type of solution appears entirely convincing, but most seem to identify a component of the response we have to painful art. To get a little further, let us take a step back.

Philosopher Gregory Currie has recently provided not a solution but a different angle on the paradox of tragedy. He frames it like this: it seems that when we watch Shakespeare's play *Othello*,[19] we want Desdemona to survive. However, we would be angry if some rogue director answered our call and staged a version of the play where she really did make it through the play.[20] Therefore, it seems as if we want two contradictory things at the same time: we want Desdemona to survive, and we want her dead.[21] Currie eventually rejects this as too "simple" a solution, because he believes that we do not really desire Desdemona to survive, but that we have only a type of *imaginative* desire for her to do so.

Some of these arguments, including Currie's reformulation, concede that we can experience several conflicting emotions at the same time, but the philosophical tradition often assumes that emotions are similar to logical statements, all of which operate on the same level and cannot contradict each other. More recent cognitive theories claim that our emotional system is more fragmented than that. Film theorist Torben Grodal argues that we are activating two parallel brain systems when watching cinema, a "global" system that understands what we are seeing as fictional and a "local" system that has more immediate emotional reactions to what is on-screen.[22] In the case of a horror film, part of the brain will assume that we are facing a genuine threat (hence we jump in our seats), while another part of the brain will understand that we are seeing a mere representation of a threat (hence we do not run from the cinema). Building on

Table 2.1
The two contradictory desires of tragedy

Immediate desire	Aesthetic desire
Desire for protagonist to succeed	Desire for protagonist to suffer as part of aesthetic experience

Grodal, Jonathan Frome has argued that we also have such multiple contradictory reactions in response to video games.[23]

The simple explanation that Currie rejected may then be on the right path. We really *do* desire two contradictory things: we want Desdemona to survive, and we want her not to survive. We maintain these two desires simultaneously, but they have different time frames: the desire for Desdemona to survive is a moment-to-moment identification with her situation, but the desire for her to die is directed at the total aesthetic experience of the play. Table 2.1 shows this as two desires with different time frames. *total aesthetic experience*

This does not solve the paradox but restates it as a balance between the short-term immediate concern for the well-being of a protagonist and the longer-term concern for the totality of an aesthetic experience. Let's then return to the paradox of failure, recalling the children's book *Liam Wins the Game, Sometimes*[24] quoted in chapter 1. The child is explicitly told that it is acceptable to feel bad about losing, but that he or she must accept losing *in order to be able to play*. In effect, Liam's book, and the general ideal of sportsmanship, describe game playing as an immediate desire for avoiding failure, combined with longer-term desires for an experience that includes failure as a necessary component in order to promote the aesthetic (or social and moral) qualities that follow. Table 2.2 illustrates how

Table 2.2
The two contradictory desires of failure

Immediate desire	Aesthetic desire
Desire to avoid failure	Desire for an experience that includes partial failure

the paradox of failure can be similarly recast as two contradictory desires.

The big difference between the paradox of tragedy and the paradox of failure is therefore that whereas the philosophy of tragedy is mostly the province of philosophers, the philosophy of failure is taught to children at an early age. Perhaps: when playing games, we are all philosophers.

The Deniability of Game Failure

The most important thing in life is not the triumph, but the fight; the essential thing is not to have won, but to have fought well.
—The Olympic Creed[25]

In chapter 1, I quoted the headline saying, "This Sport Is Stupid Anyway," following the defeat of the US soccer team in the 2010 World Cup. This is how failure in games work: it is clear that failing in a game may reflect badly upon us, but games also give us plausible deniability of responsibility: we can blame the game or luck, we can deny any actual intent to win, we can dismiss the event out of hand due to the task being *a game* (embodying the philosophy that game failure is not genuinely painful), we can deny that the game requires any skills worth having, or we can appeal to the compensatory theory of failure expressed by the Olympic Creed, in which the possible pain of failing is outweighed by the more important joy of participating.

This means that the uncertain meaning of game failure is a feature, not a bug: it allows us to take games seriously but also grants us a freedom from consequences, a freedom to deny their judgment because they are artificial designs. We could have hoped that one of the traditional explanations of the paradox of painful art would be a perfect fit for games, but the truth is rather that games are defined by the uncertain meaning of failure. We therefore have a way to save face, whenever we fail.

• To deny that we ever find game failure painful cannot be the full explanation (we often take games too seriously), but games at the same time give us the option to deny that we care about failure in this specific game session. The option of being dishonest about our feelings *by itself* makes failure easier to deal with, more socially acceptable.

• Compensatory solutions are somewhat at odds with our ability to dismiss game failure, but they make the valid point that game failure is at least partially painful (though we continue to discuss how much) but often adds a positive value to the game. The argument is also present in the different variations of sportsmanship, where failure is a price worth paying for the benefits that playing gives us.

• Finally, the solution that denies that humans exclusively seek pleasure can be seen if we interpret the Olympic Creed to be a little more austere than I previously suggested. If the creed emphasizes the fight over the triumph, the image conjured is not one of gleeful pleasure but of an intense personal struggle, performed over many years, a struggle that foregoes immediate pleasures in favor of a painful and satisfying training regimen.

[fight over the triumph] — fight over the winner — represents humanity

The paradox of failure is thus not a philosophical exercise—it is actively discussed in the conventions around game playing: we should care about winning and losing, but we should not care in a way (or at a time) that would spoil "the game," estrange our friends, or hurt other people.

It is the threat of failure that gives us something to do in the first place. It is painful for humans to feel incompetent or lacking, but games hurt us and then induce an urgency to repair our self-image. Much of the positive effect of failure comes from the fact that we can learn to escape from it, feeling more competent than we did before. This connects games to the general fact that it is enjoyable to learn something, but it also shows games as different from regular learning: we are not necessarily disappointed if we find it easy to learn to drive a car, but we are disappointed if a game is too easy. This means that failure is integral to the enjoyment of game playing in a way that it is not integral to the enjoyment of learning in general. Games are a perspective on failure and learning *as* enjoyment, or satisfaction. Of course, we can also set ourselves personal goals outside games (memorizing 1,000 digits of pi, learning to cook Thai food, and so forth), and be disappointed that it is not as much of a challenge as we thought. In such cases, we are seeing a regular learning activity *as* a game.

To play a game is to step partially through the looking glass, to cooperate with other people about being uncooperative. It is to participate in a carnival where, for a short period, our regular rules and regulations do not quite apply; where the child beats the adult and the employee is openly unhelpful to the boss; where players have elaborate explanations of how to understand and control the emotions that we experience when winning and losing, but where we do hold grudges and reveal ourselves to be sore losers nevertheless.

3   The Feeling of Failure

A few years ago, my new mobile phone came with the game *Super Real Tennis*.[1] Every time I played the game, no matter how far I progressed, each session concluded with the message shown in figure 3.1, telling me to check the "basic operation methods." These instructions told me only the most rudimentary information about the game's controls,[2] but I kept returning to them to make sure I had not missed anything. Why did the game keep telling me to read the *basic* instructions? Was I that bad a player? Were there normal and advanced instructions screens that the game would show to other, smarter players? I was rationally convinced that this was a simple design oversight, but I could not escape severe self-doubts, due to a small free tennis game on my mobile phone. 1st world problems

Some failures hurt more than others do. In chapter 1, I discussed how the term *tragedy* is usually reserved for undertakings of some magnitude: narrowly failing to win the Tour de France after years of trying qualifies as tragic, but failing to complete a small puzzle game on a train ride does not. My failing in *Super Real Tennis* was not tragic by any standard, but the more time I sank into playing the game, frustrated by my lack of progress, the worse it felt. Why was I spending all this time on a game

Figure 3.1
Super Real Tennis (Sega 2003)

that was insulting me? With all the time I had spent playing, why had I not completed it yet?

I can only admit this reluctantly. I work with video games every day, so I feel that I should be able to conquer this simple game. By admitting to my failure, I run the risk that someone will reveal my rational explanation (a developer oversight) to be wrong, showing that I really did overlook an obvious button or technique. Disassembling another version of *Super Real Tennis* revealed a screen of advanced instructions, but I do not know how these instructions would be accessed—or if they were present in the version on my phone, a Sony Ericsson K700i.

The previous chapter saw failure as a philosophical paradox of contradictory observations about human behavior, a paradox that stems from the way we are torn between an immediate desire to avoid failure and a longer-term desire for an experience that includes failure. In this chapter, I will go into further detail by examining failure through the lens of psychology. Here, failure is not a logical or philosophical problem, but a concrete phenomenon with an emotional cost that hinges in part on personality, personal beliefs, and methods of communicating failure to us.

One sports psychiatrist follows a common thread and tells us to "replace frustration with curiosity" when performing poorly—to see failure as a learning opportunity, and to play for the pure joy of the game, rather than for whatever material or psychological gains we hope to achieve by succeeding.[3] The paradox of failure thus reappears in sports literature as an attitude toward our own playing: we should care enough about winning to put in sufficient effort, but we should see learning, rather than winning, as the ultimate goal. In a similar vein, sports writers often celebrate the top player who claims to consider winning a minor detail compared to his or her love of *the game*. For example, a recent *New Yorker* profile of chess grandmaster Magnus Carlsen notes with some admiration: "Carlsen wasn't thinking about being the best, he recalled: 'I was just enjoying the game, really. I don't think I've ever really been much into setting myself these goals. It hasn't been necessary. I mean, just playing the game has been enough for me.'"[4] Of course, this is an ideal—this is not how most people behave most of the time, and it

Learning, not winning.

may be that top athletes deliberately play up to the ideal in interviews.

That said, focusing on learning objectives may well be effective, and it does stand in stark contrast to unproductive *self-defeating* behaviors: some players set themselves up for failure in order to avoid feeling that failure is a genuine measure of their skills, of who they are. Not practicing and staying up late before the game are common strategies for that. In psychological literature, self-defeating behavior is often referred to as "self-handicapping," but this term is misleading in regard to games, because there are at least four different ways in which we can play that makes it harder for us to succeed in a game, but that are not self-defeating as such:

1. Playing with defined handicaps (in games such as golf or Go) in order to balance a game.

2. Playing badly in order to keep a game interesting.

3. Playing badly in order to avoid the social consequences of succeeding.

4. Playing badly in order to explore other aspects of a game.

In these four examples, the player decides to play for a goal other than the nominal goal of a game, and the player is thus self-handicapping in regard to the game goal but is not engaging in personally self-defeating behavior. Genuine self-defeating behavior only happens when we want to succeed in a game, but we avoid practicing or taking the needed steps to actually make it happen. Self-defeating behavior is a paradoxical response to the paradox of failure: players seeking out the failure that they so desperately want to avoid. But the sports psychiatrist's advice is equally paradoxical: if you really want to win, play as if winning, and the spoils of winning, are not your primary goal.[5]

Psychological Costs

Attribution theory explains how for every event—such as a failure—we search for a cause. If we fail in a game, we consider whether to blame ourselves, the game, or some other factor. Within education, attribution theory has been used to explain the phenomenon of *learned helplessness*: viewing failure the wrong way can shake our beliefs in our own competence and *teach* us to feel helpless in the face of future challenges. Learned helplessness can be a problem in an educational setting, where the wrong kind of feedback can make a student give up entirely. One useful formulation describes three dimensions to how we deal with failure.[6] For each of these dimensions, the first option is more likely to lead to learned helplessness:

• *Internal vs. external:* attributing failure to the user *or* to the test (game).

• *Stable vs. unstable:* whether the user believes failure to be consistent *or* subject to chance or improvement.

• *Global vs. specific:* whether the user attributes failure to general inability *or* inability in this specific task.

The worst case for a student is when he or she attributes a failed mathematics test to a cause that is *internal* (my fault), *stable* (cannot be changed), and *global* (lack of intelligence in general). Since the student believes that no improvement is possible, he or she risks feeling permanently helpless in the face of future tests.

This explains why *Super Real Tennis*'s way of communicating made failure painful: by repeatedly referring to the basic operation methods, the game prompted me to attribute failure to an *internal* cause (me) and to believe that the cause was *stable* with no likelihood of improvement (since the game did not acknowledge that I was making any progress). By bringing up the basic instructions, the game implicitly signaled that I was uniquely unskilled (a possible *global* lack of skill and intelligence). A player may end up feeling quite helpless.

We generally prefer feeling more skilled than other people, but being compared negatively to other players can in some situations be motivating: the common experience in *Brain Age*[7] (figure 3.2) is to be told that our brain is older than our actual age, hence comparing us negatively to the average human. The game then promises us that we can rectify this by playing the game. *Brain Age* and *Super Real Tennis* are similar in that they give us clear hints concerning the causes of our failures. (Many games do not communicate this directly.) Both games tell us to attribute the poor result to the player (*internal* cause), but where *Super Real Tennis* suggests that the player is not improving (*stable*

Figure 3.2
Brain Age (Nintendo 2006)

cause), *Brain Age* promises that the player will improve with practice (*unstable* cause). This small difference makes failure less painful and more motivating in *Brain Age*.

The theory of learned helplessness provides a set of sensible, if slightly bland, guidelines for how games should give feedback to players: players should be told that they can improve (*unstable* cause) and that their poor performance in no way reflects on their general intelligence (*specific* cause). The internal/external dimension, however, is a little more complicated. It makes sense that we prefer *not* to feel responsible for failure—"failure is an orphan"—but the feeling of being responsible is logically the only path for us to see the possibility of improvement. How does this work in games? In the study of the prototype game discussed in chapter 2, I asked players to explain why they had failed, and then I compared their explanations to their ratings of the game's quality.[8] Figure 3.3 shows that players rated the

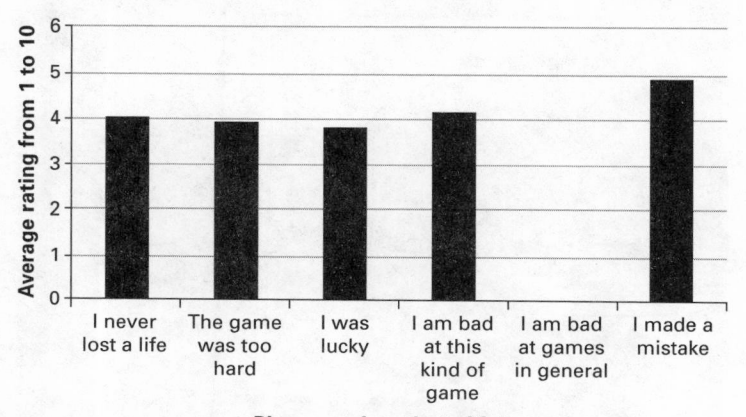

Figure 3.3
Player rating of a game as function of failure attribution

game significantly higher when they felt responsible for failure than when they did not.[9]

This marks another return of the paradox of failure: it is only through feeling responsible for failure (which we dislike) that we can feel responsible for escaping failure (which we like).

(paradox)

Abusing the Player

Based on this, we might assume that players will always dislike games that insult their skills or otherwise tell them that they will never improve (*stable* cause). However, a small group of games flaunts good manners. *Pierre: Insanity Inspired*[10] is an experimental game created in collaboration with the Singapore-MIT GAMBIT Game Lab. *Pierre* verbally insults the player with exclamations such as "This is going nowhere!" (figure 3.4).

Figure 3.4
Pierre: Insanity Inspired (GAMBIT 2009)

Responses to the game were divided between those who felt genuinely insulted and disturbed by the game's taunts (as the theory of learned helplessness would predict) and those who took it as an extra motivating challenge (as predicted by the idea that games allow transgressions of regular courtesy). Only the latter group took the insults handed out by the game as playful and motivating.

In a similar vein, the commercially successful *Portal 2* lets the player play through a series of puzzles while being taunted by the cruel computer GLaDOS. Figure 3.5 shows GLaDOS telling the player, "Remember before when I was talking about smelly garbage standing around being useless? That was a metaphor. I was actually talking about you." The high-level pattern is that games allow the transgression of many social conventions such as the general helpfulness we assume between friends and the ease of use that we expect designers to value. At the same time,

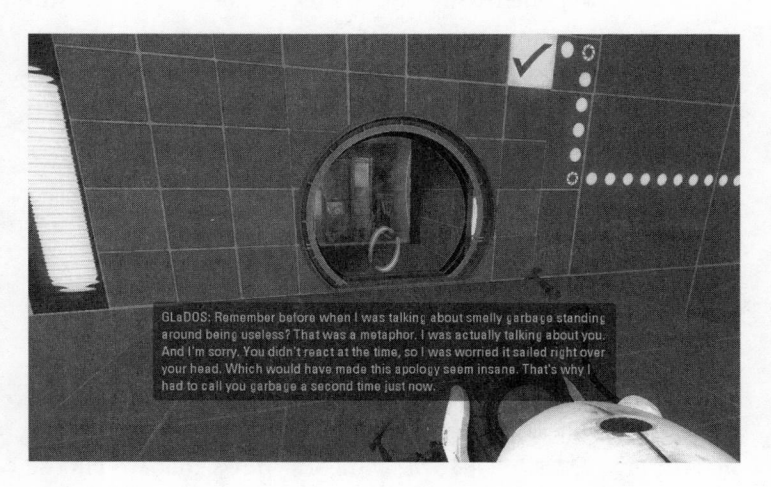

Figure 3.5
Portal 2 (Valve 2011)

this license is incomplete and often redacted: some players do feel that their loved ones should always help them, even in competitive games, and not all players accept a game that taunts them. This is the fractal nature of our reaction to games: in the big perspective, games appear to suspend the regular norms that dictate how we should be helpful and courteous toward each other. Yet, the same norms can reappear in their entirety once we examine the details of game playing. I stated that games give us license to thwart the intentions of other players, but a strategy guide to the board game *Carcassonne* warns that it is "generally frowned upon"[11] to place a piece that directly prevents another player from completing a point-scoring city or road in the game. From a mathematical point of view, there is no difference between a player gaining one point and an opponent not gaining a point (at least in a two-player game), but while the guide's author necessarily believes that a game allows a player to thwart an opponent's high-level intention of winning, he still thinks that some in-game actions are too confrontational. (It is not my experience that there is "general" agreement on this issue, but this only demonstrates that we are dealing with an often-disputed facet of game playing.)

Games as Emotional Gambling

Finnish game designer Aki Järvinen has argued that "the road to attaining goals is beset by emotions."[12] We are emotionally affected by games, and we are aware of this before we start playing. In fact, we are likely to weigh the likelihood of failure, and our possible response to it, when we decide which game to play (if any). *Mood management* theory shows that we choose entertainment in part because we want to control our mood.

We can try to escape boredom by selecting an exciting yet scary entertainment option, but we can also attempt to modify a negative mood by choosing something cheerful, such as a comedy show.[13] Given that video games generally compete with other entertainment options, video games are probably chosen in part based on their mood-managing effects.[14]

We presumably play many games because they are exciting, but will playing a particular game result in a positive or a negative mood? This is where the fundamental unpredictability of games comes in, since failure will likely result in a worse mood than success will. To play a game is to take an emotional gamble. The higher the stakes, in terms of time investment, public acknowledgment, and personal importance, the higher are the potential losses and rewards. We make very rough estimates of this gamble, factoring in the likelihood of failure along with the time investment required, the audience for our performance, and our personal investment in performing well. We are probably also not very good at doing the calculation—optimists may be unable to believe that failure is a possibility, for example. yes

While I played *Super Real Tennis*, I continued to make these emotional gambles. Having already become frustrated by the game, frustration was the basic mood that I hoped to escape by finally completing the game. I knew that the more time I sank into the game, the higher the likelihood that I would complete it, but conversely I would also experience more frustration for having spent too much time on futile attempts at the game. The greater my frustration, the larger was my motivation to escape that frustration—by playing more, which merely increased my frustration.

I was playing in a vacuum: I had never met anyone who also played the game, and information on the Internet was sparse.

My feelings changed according to how I imagined others were playing: if everyone shared my difficulties, no problem. If I were the only one who was stuck, the unpleasantness of playing grew to the point where I came close to deleting the game. Even when playing a single-player game, we still compare ourselves to others. This stems from the fact that most games are repeatable: they are also available to other players, and we will try harder if we think about how others might be performing, even if we will never know.[15] (Only when we make up games for ourselves are we free from the knowledge that others may be doing better than we are.)

For my goal of completing *Super Real Tennis*, I was not reacting optimally to failure. One psychological study distinguishes between mastery-oriented children who were likely to focus on remedies for failure and helpless children who were likely to attribute failure to lack of ability.[16] A study of video games in the classroom observed similar differences between students who saw failure as an opportunity for learning, and those who were discouraged by it.[17] I was stuck, discouraged by failure, unable to let go, but not learning as much as I should. My behavior would surely have appalled any sports psychiatrist.

In discussions of attribution theory and our emotional response to failures and successes, it can sound as if we deal with failure in a logical, linear fashion, first (1) figuring out whether it is our fault and then (2) having an emotional response. However, we are a little less honest than that. *Motivational bias* describes the fact that our emotional needs will often influence the way we make attributions. As one psychologist puts it, "The need to maintain self-esteem directly affects the attribution of task outcomes."[18] We will refrain from accepting responsibility if it is too painful. Again, the adage that "Success has many fathers, but failure is an orphan" captures this very well. And again, wanting too strongly to succeed can make us unwilling to accept responsibility for failure, and hence unwilling to improve our playing enough to actually succeed.

The Depth that Failure Reveals

We are more likely to search for causes for failure than for causes for success.[19] Whereas success can make us complacent that we have understood the system we are manipulating, failure gives the opportunity to consider why we failed (as long as we accept responsibility for failure). Failure then has the very concrete positive effect of making us see new details and depth in the game that we are playing. It is not sufficient for a game to offer interesting strategies and variations; we have to actually consider and use them. This is how games help us grow: we come away from any skill-based game changed, wiser, and possessing new skills.

Figure 3.6 illustrates the process of playing a skill-based game as a continuous cycle where a new goal is introduced; the player fails at achieving that goal, searches for ways to overcome failure, and finally succeeds. At the end of each cycle, the player has returned to the original nonlacking situation, but with new skills.[20] *player returns w/ new skills*

Players Who Optimize Too Much

The contribution of failure becomes even more clearly visible when it is absent. It is not that growth cannot happen without failure, but that failure concretely pushes us toward personal improvement, and players often *need* to be pushed because they, as game designer David Jaffe has said, are fundamentally lazy.[21]

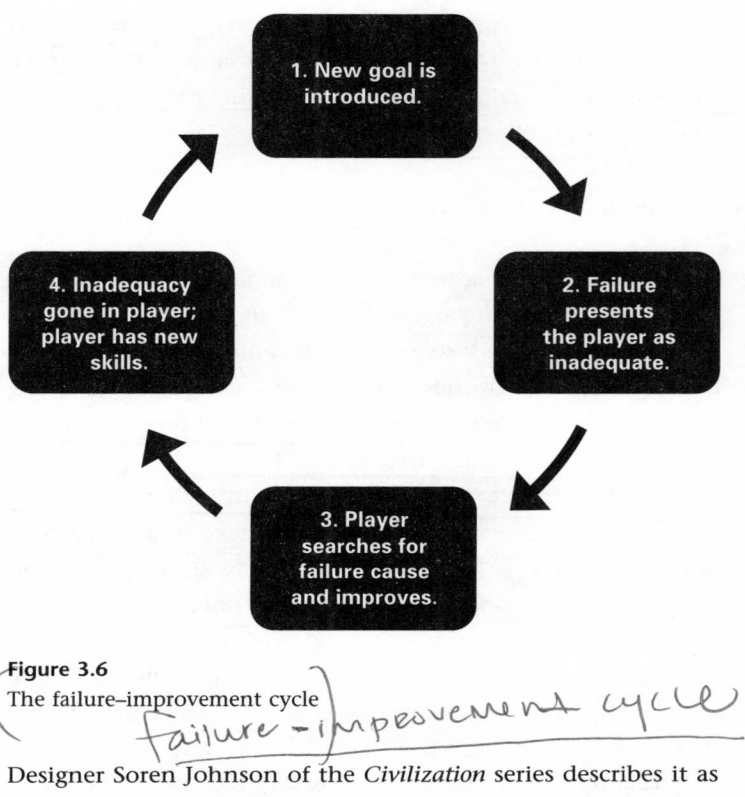

Figure 3.6
The failure–improvement cycle

Designer Soren Johnson of the *Civilization* series describes it as a general problem that players seek out the optimal path to play a game but stick to it even when they find it fundamentally uninteresting. The strategy of lumberjacking in *Civilization III*[22] (figure 3.7) is one such example:

[*Civilization III*] provides a simple example with "lumberjacking"—the practice of farming forests for infinite production. Chopping down a forest gives 10 hammers to the nearest city. However, forests can also be replanted once the appropriate tech is discovered. This set of rules encourages players to have a worker planting a forest and chopping

Figure 3.7
Civilization III (Firaxis Games 2001)

it down on every tile within their empire in order to create an endless
supply of hammers. However, the process itself is tedious and mind-
numbing, killing the fun for players who wanted to play optimally.[23]

This shows an additional layer of complexity in the psychology
of players. Michael J. Apter's *reversal theory* is a highly general-
ized view of the difference between the activities that we perform
for external goals (such as work to earn a living), and activities
that we perform for their own sake (such as playing games
or mountaineering). Apter claims that we seek low arousal in
normal goal-directed activities such as work, but high arousal,
and hence challenge and danger, in activities performed for
their own sake, such as games.[24] This is another angle on the

paradox of failure, Apter saying that games are a special type of activity in which we approach experiences that we otherwise shy away from. However, the *Civilization III* example sows doubt about the completeness of his theory: why should it be necessary for failure to force players to discover new strategies in a game? Reversal theory predicts that players actively seek out challenge and danger in a game, hence discovering failure and depth on their own. But it seems that when failure is absent, players often play it safe and do not seek the high arousal, challenge, or personal growth that we would expect.

The explanation may be that we think of single-player games as designed experiences that we expect to be correctly balanced without having to seek additional challenges ourselves. In contrast, Jonas Heide Smith has documented how players of multiplayer games frequently handicap themselves when ahead in order to maintain excitement in a game, effectively exposing themselves to failure.[25] Perhaps, multiplayer games and open sandbox games such as *Grand Theft Auto IV*[26] encourage us to undertake more challenge-seeking behavior, to seek out the depth of a game on our own volition.

For games, Apter's reversal theory must be augmented with a distinction between the *decision* to engage in a challenging activity and the *execution* of the same activity. To decide to play a linear single-player game is to decide to seek high arousal and unnecessary work. While playing the game, players tend to seek the easiest path and try to avoid failure. This matches the view of the paradox of failure as the combination of a short-term goal of avoiding failure and an aesthetic goal of engaging in an activity that includes failure. The task of the game designer is to balance these short-and long-term goals by making sure that the path of least resistance is also the most interesting one.

Least
resistance

Self-Defeating Play and Spectacular Failure

The obvious response to failure is to practice harder, but players often choose a less obvious response. *Self-defeating* behavior is a well-tested—but generally unproductive—way to avoid being measured by a task. Failure makes us search for a cause to which we can attribute our failure, but we can also influence our attributions of future failures. It is well documented that some students "prepare" for tests by staying up late, consuming alcohol, or plainly procrastinating. Should they then perform poorly, it is easy to attribute the failure not to lack of ability but to lack of sleep, being hung over, or not having studied.

Does self-defeating behavior also occur in games? A study of players of a pinball game showed that those who had neglected to practice were not as demotivated by failure as were those players who had prepared properly. Self-defeating behavior buffered players from the negative effects of failure because they could deflect the responsibility for failure (after all, they did not try all that hard).[27] Games offer many such opportunities for self-defeating behavior. We also avoid being measured by attempting challenges that are either much easier or much more difficult than would fit our abilities—the too easy challenge offers no chance of failure; the too hard challenge will not make us feel responsible for failing (since the game was obviously unfair). By avoiding the uncertainty of outcome that otherwise characterizes games, we can sidestep the feeling that we were genuinely evaluated, thereby making failure less painful.

Self-defeating students are no laughing matter, but games are a little different from studying because games make it easy for us to redefine what we consider a success. When performing a stunt in the skateboarding game *Skate 2*,[28] it is possible to fail in any number of ways (falling over, colliding with walls, finding program glitches, landing headfirst in a trashcan, etc.) Some players go as far as to share their most spectacular failures online as in figure 3.8. These players are nominally acting in self-defeating ways by not pursuing the official goal of *Skate 2*, but a more accurate description would be that they are repurposing the game by creating a new goal for themselves. This is possible because the game does not enforce its goal too strongly and hence refrains from punishing players too harshly for failure.[29] On the other hand, some of the failures posted online probably started out as earnest attempts at scoring highly in the game, but were only retrospectively redone as spectacular failures, now promoted as showing off the skill of player. In the history of video games, *SimCity*[30] was arguably the first game to explicitly support spectacular failures by allowing players to unleash various disasters (including earthquakes and monster attacks) on their creations.

Self-defeating behavior and spectacular failure are two strategies through which we can make failure feel less negative by actively seeking it out. These strategies can also be found outside games, of course, but they are more readily available for us in games because games give us a license to at least pretend that we think of game failure as unimportant. Contrast this with the self-defeating student, for whom such behavior can have serious lifelong consequences.

Mea Culpa

Attribution theory and the theory of learned helplessness explain how we comprehend failure and how emotional our stake in it can be. To play a game is to engage in an emotional

Figure 3.8
Spectacular failure in *Skate 2—Best of Reel* (Gamehelper 2009)

gamble, betting on a future emotional state at the end of the game. At the same time, much advice on how to perform to the best of our abilities will surprisingly tell us *not* to play in order to succeed but in order to learn.

We do not always behave like that. We do put in much effort to avoid experiencing unpleasant failure, to make sure that the emotional gamble of game playing pays off well, but we are not always doing this logically—by improving our skills. Sometimes we will play suboptimally to avoid feeling that failure is an actual measure of *us*. These are paradoxical answers to the paradox of failure, but they show how strongly our emotions are mixed up in our relation to failure.

Games are remarkably similar to other, more "serious" rule-based pursuits such as politics or education. As such, much of the psychology of game playing is closely related to these other activities, but it remains a defining feature of games that their tangible consequences (such as playing for money) are negotiable rather than fixed. What is not fixed, but not purely negotiable either, is how personally valuable we consider it to succeed in a given game. Our emotions toward failure also hinge on a broad and open question with existential implications: the third distinction in the theory of learned helplessness was whether we perceived our failure as local, pertaining only to the specific challenge of a specific game, or as global, a reflection of our general skills and intelligence. This is not the standard question of whether games can teach us useful skills, but the reverse: Does the fact that I failed mean that I came to the game lacking skills, intelligence, charm, or any other positive personal quality? More generally, was this failure a failure of me being who I want to be?

Logically, we should first be deciding on culpability, and then have an emotional response, but the truth is that we let our attributions be influenced by our desire to protect our self-esteem. The basic trick of learning and improvement is that we have to accept the painful answer (this is my fault, and a failure of me being who I want to be) in order to be motivated to become who we want to be. This is how each moment-to-moment attempt to avoid failure has existential significance for us.

Moment-to-Moment attempt to avoid failure has existential Significance for us.

To play a game is to engage in an emotional gamble.

the painful answer.

4 How to Fail in Video Games

In my frustration over *Patapon* (chapter 1), I thought the game had broken a promise to me. Though no promise was actually stated, I felt that I deserved a reasonable chance of crossing the desert and eventually completing the game, without having to memorize every single bit of information given to me, second-guess whether I was in time with the music, or backtrack to earlier stages in the game. Whereas chapter 2 examined failure as a general paradox, and chapter 3 dove into the psychology of failure, this chapter examines the details of how game designs make failure personal.

I thought *Patapon* was unfair, but as much as I would like to appeal to fairness, the idea of what makes a video game "fair" has changed considerably over the last few decades: UK game developer Ste Pickford has explained how when he worked on home computer games in the late 1980s, developers were "never expected to be able to complete our own games. We just presumed that some expert player out there might be good enough to get to the end."[1] Only when he began developing for consoles did publishers ask him to prove that his games could actually be played to completion. During the history of video games, the overarching trend has been toward making stronger promises to players that they will be able to overcome failure and succeed. Video games have become easier, and games that buck this trend are now lauded for their *"old-school difficulty."*[2]

Many players and developers argue that games are becoming easy in a way that detracts from their quality. One independent developer complains, "I'm not interested in making games easier, or dumber, or more boring. While Nintendo and PopCap have seen incredible financial success courting casual gamers, much of what they offer is repellent, condescending, boring, insipid, and unfair."[3] I think such a criticism overstates the case. To discuss whether a game is easy, or unfair, we should first distinguish between *failure* and *punishment*: when we say that games are becoming easier, it concerns not only our likelihood to fail at a given task but also the punishment we receive when failing.

Compare the recent big budget game *Uncharted 2*[4] (figure 4.1) to the quite challenging experimental game *Flywrench*[5] (figure

Figure 4.1
Uncharted 2 (Naughty Dog 2009)

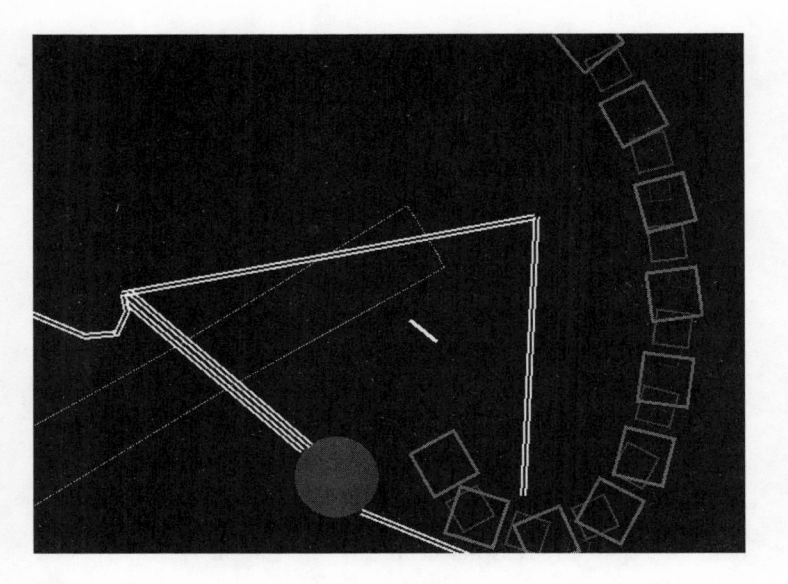

Figure 4.2
Flywrench (messhoff 2007)

4.2). In *Flywrench*, the player tends to "die" every few seconds, but at the same time, the player has infinite lives, and it only takes a second or two to recover from failure. If we think not about the number of failures but about the amount of time the player loses when failing, it becomes clear that *Flywrench* is less punishing than it would at first appear and that it gives players good opportunity to experiment. The big budget *Uncharted 2* is remarkably similar in its structure of infinite lives and closely spaced save points, dealing the player only minor setbacks for failure. In this way, the ostensibly difficult and experimental *Flywrench*, the accessible and popular *Uncharted 2*, and independent games like *Super Meat Boy*[6] with its "old-school difficulty" are all part of the same trend toward smaller punishments for failure.

In the early history of video games, the dominant way to deal with failure was the arcade model with its limited number of failures per game.[7] During the 1990s and 2000s, single-player games gradually began giving players unlimited lives. With some melancholy, one writer recently described this development as "The Slow Death of the Game Over."[8] Yet the fact that we are given infinite retries also means that we now fail more frequently than we used to, even though we often describe newer games as easy. The combination of more failures and smaller punishments adds up to more frequent opportunities for having failure force us to reconsider our strategies, to learn from our mistakes.

This was the background for my reaction to *Patapon*: had I played the game in the 1980s, I would have considered its demands perfectly reasonable and would happily have invested much time figuring out which mistakes I was making, and I would have been content to backtrack to earlier parts of the game in order to amend or collect whatever was needed. In the 2000s, I experienced this as lazy design, the result of poor user testing, or an artificially inflated difficulty curve designed only to prolong a game that was otherwise too short.

Three Kinds of Fairness, Three Paths to Success

I waited some time, then went back to *Patapon* and progressed past the desert. Though I tried to blame the game, I also felt inadequate for having at first failed to progress. By default, it is easiest to think of failure as caused by lack of skill, but in actuality a game can set up players for success and failure in at least three different ways. I failed, I think, due to a combination of lack of skills, sheer bad luck, and insufficient time investment.

We often distinguish between *games of skill* and *games of chance*, the first one tied to personal ability[9] and the second tied to luck. By convention, chess is a prime example of a game of skill, while dice games are paradigmatic games of chance. Many games, from poker to *Bejeweled*[10] are in actuality combinations of skill and chance, where the player has to navigate the probabilities of the game in order to succeed. But this is not the whole picture. Consider *World of Warcraft*[11] shown in figure 4.3. This game belongs to the genre of Massively Multiplayer Online Role-Playing Games (MMORPGs) and requires players to play for weeks and months. A large part of that time is spent performing repetitive battles with computer-controlled creatures in order to progress in the game (affectionately known as "the grind").

Figure 4.3
World of Warcraft (Blizzard Entertainment 2004)

Game designer Naomi Clark suggests that this is a third type of game, a game of *labor*, in which players are rewarded for performing routine tasks, rather than for their skill or for their luck.[12] Games then promise players the possibility of success through three different kinds of fairness or three different paths: *skill*, *chance*, and *labor*. The three paths are in practice often combined, but let me first examine them separately.

Skill

Learning through failure.

The promise of success through skill is the least controversial of the three paths. As players, we come to a game with a repertoire of skills that we try to apply to the problem at hand.[13] We can continually improve our skills, and whenever we fail, we have the chance to reconsider our strategies, to recalibrate or expand our skillset. Success through skill is hence the path to success that is most closely tied to the experience of learning through failure, and to the larger claim that games are fundamentally *about* learning.[14] Skill is therefore also closely aligned with the idea of game playing as a source of personal improvement, both in the moral sense associated with sportsmanship and in the broader sense that games can be used for education and for corporate or societal good. When Benjamin Franklin, as quoted in chapter 1, claimed that chess teaches valuable life skills, he was focusing on its qualities as a game of skill. When we fail in a game of skill, we are therefore marked as deficient in a straightforward way: as lacking the skills required to play the game.

Chance

The distinction among skill, chance, and labor has concrete legal implications. Of the three, chance has historically been the most controversial, mostly for its association with gambling (even in games where no exchange of money took place). Pinball, banned in New York City from the 1930s on, was finally made legal in 1976 when an expert player demonstrated before the court that pinball was, in fact, a game of skill.[15] More recently, the Supreme Court of Denmark decided that poker was a game of chance rather than skill and that the Danish Society of Poker Players was therefore not allowed to hold tournaments with cash prizes.[16] The underlying flawed assumption is that games are *either* skill or chance, when they in fact often combine skill, chance, and labor.[17] Chance is also historically used in divination practices and is occasionally considered the province of a divine power (since chance defies our regular understanding of causality). One devout Christian told me that he considered games of chance problematic because the player directly petitions the Christian God to intervene on his or her behalf. Failing in a game of chance therefore marks us in a different way than failing in a game of skill does: as being on poor terms with the gods (or *Fortuna*), or as simply *unlucky*, which is still a personal trait that we would rather not have.

Fortuna's Wheel

Labor

Labor is a relatively new type of path to success that has become controversial in relation to the currently popular social game *FarmVille*[18] (figure 4.4). For example, developer Chris Hecker denounces the game as encouraging "junkie behavior," unlike other games that he prefers: "When you're playing *Counter-Strike* or even just throwing a Frisbee, the thing you're doing is fun in itself. In Zynga [makers of *FarmVille*] games, you're just trying to get more stuff. You're caught up in this junkie behavior, and you have to keep upping the dose."[19] Hecker's criticism is leveled at the *grind* of the game, through which players gain new powers

Figure 4.4
FarmVille (Zynga 2009)

by performing the relatively trivial actions of planting and harvesting. Once the player has gained new powers, these in turn make it easier to plant, harvest, and gain new powers. This type of positive feedback loop is in no way unique to *FarmVille*—it can be found in any number of games, from checkers (having doubled makes it easier to perform the next doubling) to *Counter-Strike*[20] (winning a round gives extra purchasing power to the winning team). As I interpret it, Hecker is criticizing *FarmVille* for being a game of labor exactly because its fundamental actions such as seeding and harvesting are both easy and repetitive. This game can be extrapolated to a general three-part definition of a *game of labor*:

1. The game is played in a series of small game sessions, distributed over time.

2. Players gradually accumulate more abilities (and items) that give them new powers, which are persistent between game sessions.

3. The actions needed to accumulate abilities are mostly trivial and rarely end in failure.

It follows that any criticism of *FarmVille* for being a game of labor, with the player "getting more stuff," could just as well be directed at the entire genre of role-playing games, including the aforementioned *World of Warcraft*. Again, players play a series of individual sessions, gradually accumulate abilities and items over time, and do so by performing a long series of battles and missions, most of which become trivially easy as the player's character gains new powers. The juxtaposition of these two games may seem surprising given that they are associated with entirely different aspects of game culture: *FarmVille* is generally understood as a casual, social game for a broad audience, while *World of Warcraft* is understood as an intense experience for the more dedicated video game player. However, both are games built around labor, more similar than their fans would probably like to admit. Video game emotion theorist Nicole Lazzaro has described the two games as "separated at birth."[21] The truth of the matter is that they share a heritage—namely, the point and inventory system of pen and paper role-playing games.

Figure 4.5 shows the game *Statbuilder*,[22] in which players simply have to press the large central button repeatedly in order to progress in the game—a parody of the trivial core activity and time-dependent progress of many role-playing games. Since such a game of pure labor contains resources that trivially grow

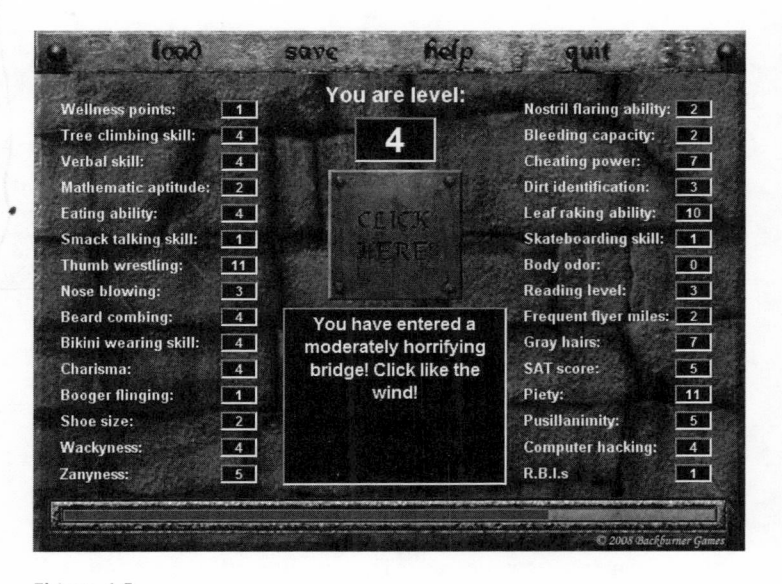

Figure 4.5
Statbuilder (Backburner games 2008)

or improve over time, it can—controversially—be completed
without the player actually learning anything.[23] Of course,
games of pure labor are rare, and *FarmVille* and *World of Warcraft*
are not based purely on labor. The *FarmVille* player needs to
become familiar with new crops, machines, and the time planning
strategies required to use them properly. The *World of Warcraft*
player needs to develop a similar (though more complex) understanding of higher-level equipment and strategies and must be
able to navigate and communicate within the social world of
the game. It is easy to determine whether a game is based purely
on labor: if a novice player can perform well when dropped in
at a late stage or level, the game in question is a game of pure
labor. Thus, *Statbuilder* as such is a pure game of labor, but
FarmVille and *World of Warcraft* are only partially based on labor.

Failure in a game of labor has a very different personal meaning from failure in a game of skill and/or chance. We may intuitively think of failure and success as mutually defined opposites, but because games of labor are played over a long time through a series of smaller game sessions that gradually propel players through the game, a game of pure labor (such as *Statbuilder*) may not even allow a player to fail. Failure, in such a case, is rather not-having-succeeded-yet and can be down-played as the fact that a player just has not gotten around to completing the next series of quests, or planting the next set of crops. Lack of success in a game of labor therefore does not mark us as lacking in skill or luck, but at worst as someone lazy (or too busy). For those who are afraid of failure, this is close to an ideal state. For those who think of games as personal struggles for improvement, games of labor are anathema.

The Politics of Fairness

We can ask which path to success is more *fair*, but this reveals that fairness has several possible meanings. Games of chance are "fair" in that they promise all players an equal chance of winning; games of skill are "fair" in that they justly reward personal skills; games of labor are "fair" in that they promise equal outcomes according to time invested. These notions of fairness extend between multi- and single-player games. Figure 4.6 shows the warning message for the "nightmare" skill level in *Doom*:[24] "This skill level isn't even remotely fair." In this case, not being *fair* refers to the near-absence of a (skill-based) path to success. This in turn serves as an extra motivator for players willing to attempt to complete it anyway, and *Doom* here appeals to much of video game culture's valuation of skill above everything else.

This set of values can also be seen in developer David Sirlin's criticism of *World of Warcraft*. Sirlin argues that the game "teaches

Figure 4.6
Doom (id Software 1993): What does "unfair" mean?

the wrong things" by downplaying skill in favor of time investment (i.e., labor), thereby telling players the following:

**1. Investing a lot of time in something is worth more than actual
skill.** If you invest more time than someone else, you "deserve" rewards.
People who invest less time "do not deserve" rewards. This is an absurd
lesson that has no connection to anything I do in the real world. The
user interface artist we have at work can create 10 times more value
than an artist of average skill, even if the lesser artist works way, way
more hours. The same is true of our star programmer. The very idea that
time > skill is alien.[25]

Sirlin's criticism hinges on the assumption that games should
work like *real life* and that *World of Warcraft* is therefore flawed
by being different from real life. This is surely too simple an
argument given that games do not necessarily claim to represent
the regular world accurately. The bigger picture is that many

popular games—including games revered in traditional video game culture—reward players for chance and labor in addition to skill. Though chance and labor are often denounced, they are integral aspects of many games across genres. At the same time, each path openly corresponds to a political ideal.[26]

• *Skill* is a meritocracy that rewards according to skill and accepts the subsequent inequalities among players.

• *Chance* has a fundamental duality between repeated games, where all participants end up benefiting equally (due to chance leveling out over time), and single games, where goods are distributed unequally. Chance is egalitarian when considered over multiple game sessions (equal distribution), but chance within a single game session is libertarian (complete acceptance of unequal distribution, even when due to chance).

• *Labor* corresponds to the Protestant work ethic as described by sociologist Max Weber, where the investment of effort is the path to salvation.[27] (Here, I add the caveat that a work ethic is hardly a Protestant invention.) This is egalitarian in the sense that it rewards only for effort, not for skill.

A recent study of Norwegian schoolchildren in fifth grade through high school revealed that our idea of fairness evolves as we grow up. In groups, the children were asked to perform an income-generating task, after which one participant was asked to distribute the income among group members. The younger children in the study were most likely to consider an equal distribution fair, but older children began to accept unequal distribution based on skill and effort.[28] However, each age group had its share of libertarians who did not believe in any form of redistribution, even when income had originally been distributed according to chance.

Skill, chance, and labor are the three primary paths to success that games offer, but they are not the only ones. Outside the "pure" game itself, recent innovations include free-to-play games with microtransactions and real money trade of virtual items, allowing players to purchase special powers or advantages. This also takes place in *World of Warcraft* and *FarmVille*. Again, we can discuss whether these are fair (leveling the playing field for those with little time on their hands) or not (shifting the focus away from the skill, chance, and labor of the game).[29]

Though I have just given political labels to skill, chance, and labor, it does *not* follow that each type of fairness is a statement of support to a specific ideology. Ian Bogost[30] and Mary Flanagan[31] among others have forcefully argued that games have built-in values. However—and this is the general problem of interpretation— how can we distinguish between a game that, for example, celebrates economic inequality and one that warns against it? In fact, this is the case with *Monopoly*,[32] one of the world's most played board games. Growing up in the 1970s, I witnessed many adults distancing themselves from the game's assumed celebration of financial ruthlessness (though the same adults were obviously enjoying the game). In one of the more amusing ironies of game history, *Monopoly* was in fact a copy of Elizabeth J. Magie's 1903 *The Landlord's Game*.[33] That game had been designed with the didactic goal of *warning* against the evils of monopoly ownership of land and promoting a single tax system: "The object of this game is not only to afford amusement to players, but to illustrate to them how, under the present or prevailing system to land tenure, the landlord has an advantage over other enterprisers and also how the single tax would discourage land speculation."[34] This means that although skill, chance, and labor map quite directly to political ideals, we cannot conclude that a game is in favor of the political ideal it may

seem to embody. If we return to the Olympic Creed, it states
that "the essential thing is not to have won, but to have fought
well."[35] The creed argues that although the game has just ranked
the players according to their skills, we should not take this
ranking at face value, but see the players as equals instead. This
is clearly a variation on the paradox of painful art: Why do we
in art seek out emotions that we regularly shy away from? And
then why do we engage in systems that are based on beliefs that
we disagree with? Part of the answer is that we understand art
as a space for exploration of possibilities. We will read novels
about war with the knowledge that this is not an endorsement
of war. And we will play games with different types of fairness,
aware that this does not amount to an endorsement of that type
of fairness as a general ideal.

Same w/ film / Books 7 — not fair to say for everyone

Completable, Transient, and Improvement Goals

When we fail in a game, we are told that we are flawed and
inadequate. Some failures brand us for life, while others are
quickly remedied. If we fail to protect our player character in
Mass Effect 2[36] (figure 4.7), we will be offered an opportunity to
retry, and the effective cost to the player is simply that we have
lost time, postponing the completion of the game.[37] Compare
this to when we fail to solve a Solitaire (Patience) card game,
or when we lose a multiplayer match of *Super Street Fighter IV*[38]
(figure 4.8). In Solitaire or a multiplayer competitive game, we
have not postponed any completion of the game, since completion is not possible in the first place. Rather failure is tied to a
specific game session, and no time invested can make up for it
after the game is over. Finally, consider what happens when we
fail to improve our high score in *Geometry Wars* (figure 4.9). Like
other games following the arcade model, and like any other

Figure 4.7
Failure to complete—*Mass Effect 2* (BioWare 2010)

Figure 4.8
Failure to win this game—*Super Street Fighter IV* (Dimp/Capcom 2010)

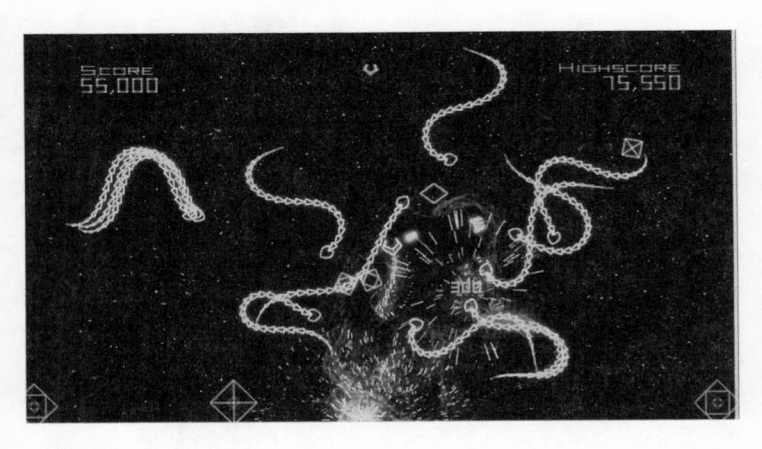

Figure 4.9
Failure to improve—*Geometry Wars* (Bizzare Creations 2005)

game we are playing with the goal of improving our personal
best, failure here is a temporary failure to improve our previous
personal record.

This is to say that failures reflect on us and have different
shelf lives, depending on the *goal type* of the game:

• *Completable goal:* *Mass Effect 2* is a mostly linear game that can be
completed once and for all. Once we begin playing, we become someone
who has-not-completed-the-game-yet. Once we complete the game, we
will always be someone who has completed *Mass Effect 2*.

• *Transient goal:* Winning a match in *Super Street Fighter IV* only means
that we have won *this* match: this is a *transient goal*, tied to that specific
game session of *Super Street Fighter IV*. This also applies to Solitaire card
games, where the goal is not to complete Solitaire once and for all but
to solve the specific set of randomized cards of *that* round of Solitaire.

• *Improvement goal:* As a compromise between completable and transient goals, the arcade game on one hand has a completable goal (beat
your personal best), but once it is achieved that goal is immediately
replaced with the goal of beating the *new* personal best. Such improvement

goals concern our ongoing personal struggles for improvement, and can by definition never be reached.[39]

Games tend to encourage us to play for a specific type of goal, but it is also possible for us to incorporate additional personal goals in a game. We can play tennis for the personal goal of winning at least one match against a rival, or we can play Solitaire to solve a specific variant for the first time. In such cases, we are still playing the nominal game with a transient goal, but we have added a completable goal on top, a goal we can reach permanently or continue to fail to reach. Part of the function of the achievement system on Xbox 360 is to institutionalize this practice. For example, the "Completely Unstoppable!" achievement in *Bomberman Live*[40] awards the player 10 gamer points for winning five consecutive multiplayer games: this adds a completable goal on top of the transient goal of each individual game session and thereby creates a general record of the player's progress across single matches.

Completable games encourage a kind of exhaustion whereby the player is meant to stop playing the main game once it has been completed. The most common exception is the practice of *speed runs*, where the player incorporates the new goal of finishing a completable game as fast as possible, hence adding transient and improvement goals ("how fast can I complete the game this time" / "can I beat my personal best") to a completable goal. Another variant on this theme is the *permadeath* play through, where the player decides to play (for example) *Far Cry 2*[41] with only one life, promising to stop playing at the moment of death.[42] Both *speed runs* and *permadeath* are in practice often public performances, documented and shared.

Failures against completable, transient, or improvement goals have different existential implications: while working toward a completable goal, we are permanently inscribed with a deficiency, and reaching the goal removes that deficiency, perhaps also removing the desire to play again. On the other hand, we can never make up for failure against a transient goal (since a lost match will always be lost), whereas an improvement goal is a continued process of personal progress. Each type of goal makes failure personal in a different way and integrates a game into our life in its own way.

Set Up for Failure

Table 4.1 shows how the three goal types combine with the three paths to success, producing a matrix of the ways in which games set players up for failure and success. To say that a game *has* a specific goal type is to say that the design encourages players to play for *that* goal. *Mass Effect 2* encourages players to complete it once and for all. *Space Invaders*[43] encourages players to beat their previous high score, while *FarmVille* lets them continually expand their farm. *World of Warcraft* is interesting in that it caters well to all three goal types: it can be played for the goal of reaching the current maximum level, but it is also possible to play with the improvement goal of acquiring ever more points, possessions, and higher social status, and it is common to play many characters to the maximum level, making it into a game of transient goals, to be reached multiple times.

Games have become easier, and therefore we fail more: it is true that video games are becoming easier overall, but primarily in the sense that they are easier to complete because they deal smaller punishments for failure than before. Within that trend, failure is actually becoming more common, with infinite retries and smaller punishments lowering the cost of failure, as

Table 4.1
Three goal types and three types of fairness

Goals/Paths		Skill
Completable goal (Complete once)	*Genre*	Single-player completable action and adventure games
	Example	*Patapon* *Limbo* *BioShock* *Far Cry 2*
	As player-set goal	Personal goal in transient game of skill: winning against a player at least once
Transient goal (Complete many times)	*Genre*	Competitive multiplayer games
	Example	Tennis *StarCraft II*
	As player-set goal	Taking turns playing a single-player arcade game to see who scores highest
Improvement goal (Goal of continuous personal improvement)	*Genre*	Arcade games
	Example	*Geometry Wars* *Space Invaders*
	As player-set goal	Speed run of *BioShock* and other skill-based completable games

Chance	Labor
?	Single-player role-playing games
	Dragon Age *World of Warcraft* as reaching current maximum level
Personal goal in transient game of chance: solving a Solitaire variant at least once	Reaching maximum level in a free-to-play game without paying
Games of chance in general	?
Solitaire Roulette Rock-paper-scissors	*World of Warcraft* as reaching maximum level through multiple characters
Impromptu games of chance: who draws the highest card?	Speed run of game in labor-based role-playing game
?	Non-completable linear games of labor *FarmVille* *World of Warcraft* as maximizing experience points, acquiring money and items
Personal best in transient game of chance: personal best in *Yahtzee*	Personal best speed run in labor-based role-playing game

measured in time. Hence we spend more energy thinking about *why* we failed, *what* we can do about it, and *how* it reflects on us personally.

A tired cliché of sports announcers and game promoters declares "This time it's personal!" This is misleading, because games are *always* personal; only game designs have different ways of making a game personal. Skill, labor, and chance make us feel deficient in different ways when we fail. Transient, improvement, and completable goals distribute our flaws, our failures, and successes in different ways across our lifetimes.

5 Fictional Failure

In a game where we control a character, any success that we achieve will almost always result in the fictional protagonist achieving success. When we fail, the protagonist almost always fails as well. In games with no single protagonist, our success or failure is similarly matched by success or failure for the city, society, or world whose interests we have been protecting.

Figures 5.1 and 5.2 show how the mirroring of player and protagonist goals means that the mood of the player is also mirrored by the mood of a fictional protagonist. When the player is happy to have completed the game, the fictional protagonist tends to be equally happy because the protagonist has also fulfilled his or her personal goals; when the player fails, both player and protagonist are unhappy.

This mirroring of player and protagonist moods is nearly universal, but does it have to be like this? Janet Murray's 1996 book *Hamlet on the Holodeck*[1] examines an episode of the *Star Trek: Voyager* television series, in which Captain Kathryn Janeway participates in a virtual reality program of a Victorian novel, falls in love with the fictional Lord Burleigh, and eventually decides to delete the program. Murray uses this to describe an appealing vision of a future interactive drama, in which the user

Figure 5.1
Success! The successful player and the successful protagonist

Figure 5.2
Failure! The unsuccessful player and the unsuccessful protagonist

Figure 5.3
A tragic ending! The successful player and the unsuccessful protagonist

is fully immersed in a virtual reality world, engages naturally with characters, and experiences love, pain, and the whole range of emotions that we expect from literature, theater, and cinema. If we follow Murray and imagine a game with a tragic ending, we would expect the mood of the player and the protagonist to be inverted as in figure 5.3, where the player is happy to have successfully overcome failure and played the game to completion, but where (this being a tragedy) the protagonist is correspondingly *un*happy. As the illustration shows, such a game has a counterintuitive disconnect between the enjoyment of our accomplishment and our empathy with the plight of the protagonist. If we knowingly play for a tragic ending, our priorities as players may not be aligned with the interests of the protagonist as much as with the completion of the story arc of the game.[2] In addition, this setup shows a disconnect between the personal flaw that makes it necessary for a tragic protagonist to meet a tragic end, and the player who in completing the game has managed to *overcome* personal flaws.[3]

Of course, many games feature protagonists with transgressive agendas—criminals, soldiers, and so on. It is common for games to let us work toward goals that we fundamentally consider abhorrent but such goals can be exciting in video game form. Video games have traditionally been concerned with themes of negative valence (i.e., scary or unpleasant) that we would avoid in real life (war, violence), but such themes also arouse us to act in order to protect a protagonist from existential threats.[4] (The recent rise of "casual" games for a broad audience has prompted the appearance of many games with positive emotional valence such as raising a garden or running a restaurant—activities that players actually aspire to, outside games.[5]) Though some naïve critics have assumed otherwise, the player of a video game does not automatically endorse the events in the game. The case is rather that game goals tend to reflect the interests of the protagonist, regardless of whether we find the actions of the protagonist morally defensible.[6] The question of tragedy in games is therefore not whether we automatically endorse the actions of a protagonist (we don't), but whether we are willing to work for a goal that contradicts the interests of the protagonist. (I am talking here mostly about video games—the role of tragedy in less strictly rule-based game forms such as experimental role-playing games would be a separate study.)

In 2001, fiction theorist Marie-Laure Ryan argued against the idea of game tragedy by saying: "Interactors would have to be out of their mind—literally and metaphorically—to want to submit themselves to the fate of a heroine who commits suicide as the result of a love affair turned bad, like Emma Bovary or Anna Karenina. Any attempt to turn empathy, which relies on mental simulation, into first-person, genuinely felt emotion would in the vast majority of cases trespass the fragile boundary that separates pleasure from pain."[7] We are comfortable with tragic stories and unhappy endings, but Ryan argues that it is the distance from the viewer to a fictional tragedy that converts negative emotions into something pleasurable (a *conversionary* argument as discussed in chapter 2). For Ryan, it follows that the first-person quality of a game would make the negative emotions of tragedy into straightforwardly unpleasant emotions.

Building on Ryan, I similarly argued that tragedy in games was impossible because "the goal in the fictional world must mimic the player's real-world situation by being emotionally positive as well."[8] At the very least, the promotional material for a game based on Shakespeare's play *Othello* would sound absurd: "Othello—the game! Get tricked into believing that your wife is unfaithful to you! Murder her, then recognize her innocence and commit suicide in shame!"[9] Murray's defense of interactive tragedy is that even in the Holodeck, it is safe to experience tragedy because we can "shut the book," turn off the experience.[10] Murray believes that this ability to control and exit the experience renders painful emotions relatively harmless, and that our ability to influence the game world does not fundamentally alter that equation. This question, of whether tragedy is possible in video games, can probably not be resolved entirely in the abstract, so let me move on to some examples.

Suicide Games

Ryan does not state that anything physically prevents a game designer from creating a tragic game, only that nobody in his or her right mind would want to play an *Anna Karenina*[11] game. In Tolstoy's novel, Anna Karenina engages in a long-time affair with Count Vronsky, makes the painful decision to leave her husband and son, is publicly humiliated, comes to believe that she and Vronsky are drifting apart, and finally commits suicide by throwing herself under a train. For a game to live up fully to this example, I believe it would have to have three characteristics:

1. The protagonist undergoes many painful experiences.

2. The player is aware that the goal of the game is to commit suicide.

3. The player exerts effort in order to commit suicide.

To test the question empirically, I collaborated with Albert Dang and Kan Yang Li who created *The Suicide Game*,[12] in which players have to commit suicide. In the game, two players work together to move the protagonist around a room (figure 5.4). To commit suicide successfully, the players must drink poison and stab themselves with the objects distributed throughout

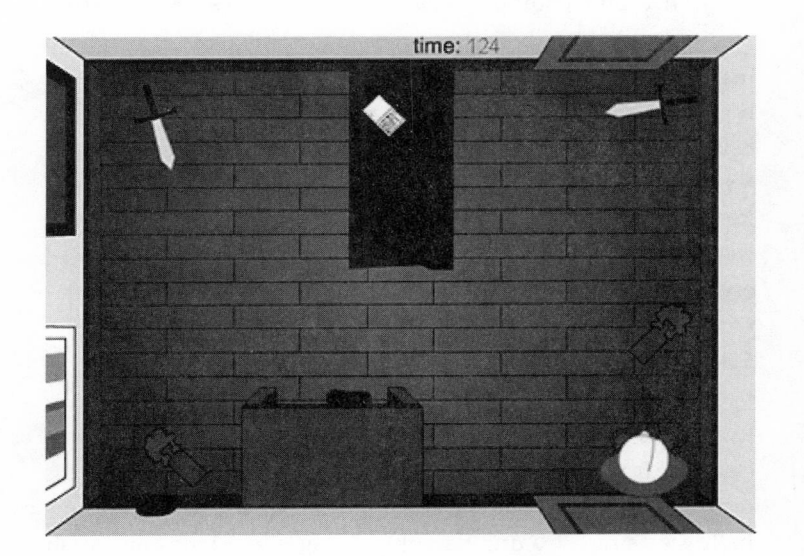

Figure 5.4
The Suicide Game (Dang, Li 2006)

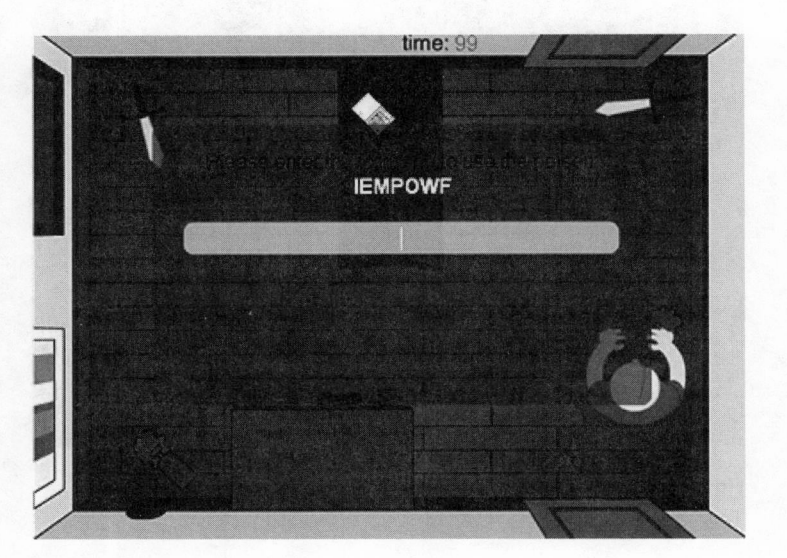

Figure 5.5
Typing a code word to drink poison

the room. Drinking poison or stabbing yourself is performed by going to the object and typing in the text string the game presents (figure 5.5). Failing to complete the game goal within the allocated time frame produces a message that the player has failed but the player character has survived (figure 5.6). Conversely, completing the game leads to the death of the player character (figure 5.7). These are exactly the juxtapositions I discussed earlier, where the success of the player is linked to the suffering of the protagonist.

▶The suicide game is playable at http://www.jesperjuul.net/ text/suicidegame

We observed some test subjects play the game, and while they did describe the game as somewhat shocking, the cartoony

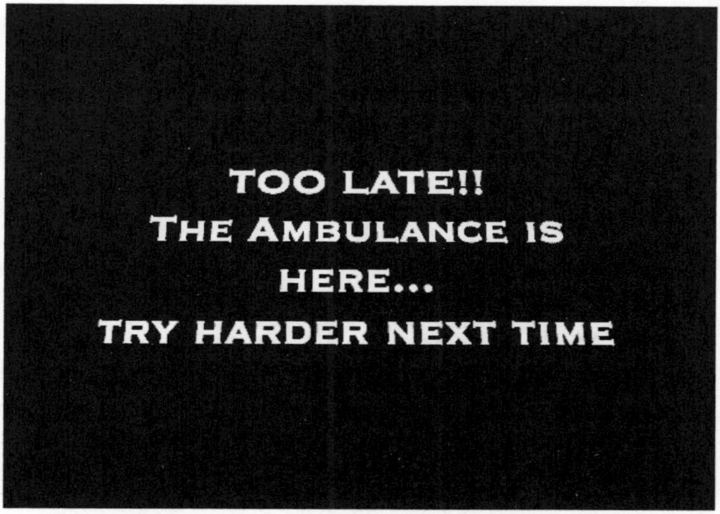

TOO LATE!!
THE AMBULANCE IS
HERE...
TRY HARDER NEXT TIME

Figure 5.6
To survive is to fail

Figure 5.7
To die is to succeed

quality of the graphics and the ironic quality of the text seemed to mitigate some of the weight of the subject matter. One player volunteered the information that a friend of hers had killed herself, but stressed "this is just a game." We interpreted this to mean both that the game had made the player associate to the personal experience and that the player believed it customary not to "take games seriously." Another player expressed what we interpreted as a joyful experience from the transgressive subject matter of the game: "This is awesome. You guys are sick."

All players reacted to the suicide game as a departure from the games they knew. They were at least tacitly aware of the fact that games generally involve the player protecting the well-being of the player character. During play, when players focused on the concrete task of coordinating their movements to navigate the playfield, the theme of the game seemed to fall into the background compared to the task of simply performing well in the game. Only when players failed or succeeded did they become aware of the goal of the game again.

Our experiment showed that the question of game tragedy is not either/or. The graphics and the tone of the game appear to make the act of self-destruction less disconcerting than it would otherwise have been. Setup, presentation, and gameplay strongly influence our experience of a tragic ending. In addition, the fact that the protagonist in this game positively *wants* to die for reasons unknown to us (as opposed to not wanting to die, or wanting to die to escape a horrible situation) aligns the interests of the player and protagonist more closely than they would be in the hypothetical *Anna Karenina* game.

Suicide and self-destruction are not entirely unheard of in commercial games, but they tend to appear in less direct form than in *The Suicide Game* and *Anna Karenina* examples. Figure 5.8

Figure 5.8
Burnout Paradise (Criterion Games 2009)

displays the "Showtime" mode of *Burnout Paradise*,[13] in which the player has to race a vehicle into traffic in order to cause maximum damage. This experience of self-destruction has an unpleasant aspect to it, but the game presents no human characters, and furthermore restarts immediately after a crash with no cost to the player, hence deemphasizing any human suffering caused.

Burnout Paradise is part of a small trend that does not involve the long-time suffering of the protagonist, but rather fascinates through the immediate joyful discomfort of witnessing (bodily) destruction. Other examples include *Stair Dismount*[14] (figure 5.9) and *Super Meat Boy*[15] (figure 5.10) as well as *Limbo* and *Super Monkey Ball* shown in figure 1.6.

The otherwise conventional *God of War*[16] shown in figure 5.11 is the reversed model of *Burnout Paradise*. Unlike the latter game, *God of War* does feature a protagonist, Kratos, who is seen trying to commit suicide in the initial cutscene. By playing

Figure 5.9
Stair Dismount (Secret Exit 2009)

Figure 5.10
Super Meat Boy (Team Meat 2010)

Figure 5.11
God of War: Kratos attempts suicide (SCE Studios 2005)

through the game, we learn of his reasons for doing so—he has been tricked into killing his family. However, in contrast to *Burnout Paradise*, the player does not control Kratos when he tries to kill himself by jumping off mount Olympus. Unsurprisingly, the box for *God of War* does *not* ask the player to "commit suicide" but rather, "Defeat Ares, the God of War." While the fictional content of this game is closer to what we expect from a tragedy, we are still only asked to perform the actions that *led* the protagonist to attempt suicide.

[SPOILER ALERT: The following discussion reveals the ending of *Red Dead Redemption*.]

Figure 5.12 and figure 5.13 show *Red Dead Redemption*,[17] which can only be completed by letting the protagonist die. The overall plot of the game is that John Marston has lived a life of crime that he has tried to leave behind. During the central part of the game, some lawmen holding his family hostage blackmail him into completing a number of missions of questionable moral value. Once that is done, Marston is reunited with his family and the player goes through a number of mundane (and quite boring) missions involving rounding up cattle, scaring away crows, and so on. Finally, the family is attacked by his earlier tormenters. Throughout the game, and during most of the attack, the player is required to "retry from checkpoint" every time the protagonist dies. To complete this final mission, Marston has to help his wife and son escape from the attackers. If the player fails before the family has escaped, the player is asked to retry as usual (figure 5.12), but after they have escaped, the protagonist faces an overwhelming opposition and is shot dead (figure 5.13). This time, the player is not asked to retry, and for the (brief) remainder of the game, the player controls Marston's son Jack instead. Asking the player to sacrifice the

Figure 5.12
Red Dead Redemption main game (Rockstar Games 2010)—dying leads to retrying

Figure 5.12
(continued)

protagonist for a greater good is unusual for video games yet is conventional as storytelling, with the stereotypically male hero protecting, or failing to protect, or hurting, his family. This does answer one of the logical problems of combining games with storytelling in general and tragedy in particular: inevitability. Mandel's definition of tragedy described how the protagonist must meet his or her end "necessarily and inevitably." The game unconventionally denies the player choice at a crucial point in order to create a conventional tragic ending.

By focusing more closely on the motivations of the protagonist, the content of *Red Dead Redemption* is closer to the *Anna Karenina* example than either *Burnout Paradise* or *God of War*. However, John Marston dies for a cause, where *Anna Karenina* kills herself in despair. Furthermore, *Red Dead Redemption* players are specifically not told that the protagonist will die if they complete the game, and they are powerless spectators to his murder.

Really?
I believe
she does die for a cause —
thoughts?

Figure 5.13
Red Dead Redemption tragic passage—protagonist must die to allow
player progress

Figure 5.13
(continued)

The response from players has been divided, with some outraged that they unusually cannot save the main character, others accepting it for the sense it makes within the story of the game.[18]

While the *Suicide Game* and these commercially successful games involve the destruction of the protagonist, they each lack at least one element from the *Anna Karenina* example: the *Suicide Game* does not present the psychological reasons for the protagonist wanting to commit suicide, and it presents itself in a somewhat ironic tone; *Burnout Paradise* does not feature a protagonist for us to identify with; *God of War* does not ask us to exert effort in order to attempt the suicide; *Red Dead Redemption* does not inform us that the protagonist will eventually die and we are only spectators to his death. While games can involve the physical suffering of the protagonist to a larger extent than assumed, commercial video game designers have so far shied away from the full suicidal self-destruction of *Anna Karenina*.

Complicity and Deception

Physical self-destruction is one variation of fictional tragedy in video games. Another type of tragedy focuses on psychological suffering: the protagonist survives but feels guilty about having caused suffering. This may sound like another contradiction at first—players still seem unlikely to do something that they do not want to! However, there is the additional possibility that we already saw prefigured in *Red Dead Redemption*: deception.

Brenda Brathwaite's board game *Train*[19] (figure 5.14) at first appears to be a simple multiplayer strategy game in which players have to pack some human figures into boxcars as efficiently as possible. Once the first player manages to get a boxcar

Figure 5.14
The feeling of complicity in *Train* (Brathwaite 2009). Photo by Geoffrey Long.

to the other end of the track, he or she must draw a card from the (face down) destination pile. All cards are names of Nazi concentration camps: *Auschwitz*, *Dachau*, etc. At this point, the game tends to stop. In the absence of prior knowledge of the game, most players fail to realize that they are participating in the Holocaust before the card is turned. Brathwaite has described the experience as one of *complicity*:[20] players suddenly realize that they have been working toward an abhorrent goal. As it turns out, this use of deception and revelation opens up a whole range of new experiences, where the discomfort of having worked for something unpleasant turns out to be a strong emotional device unique to games. The experience is not one of trivialization, but of feeling painfully involved in an event in a way we do not experience in merely fictional representations such as cinema or literature.

Shadow of the Colossus[21] (figure 5.15) provides a different take on complicity: in this case, we are told that in order to save the girl Mono, we have to kill a series of colossi. These colossi turn

Figure 5.15
Shadow of the Colossus (Team Ico 2005)

out to be large, slow-moving, melancholy creatures that inspire pity rather than fear. Although the girl is eventually saved, it becomes clear during the game that the colossi are actually innocent creatures, and the protagonist is eventually punished for having killed them. Whereas the deception in *Train* comes directly from the game designer who does not disclose the setting of the game, the deception in *Shadow of the Colossus* comes from in-game characters. In both cases, the result is one of feeling complicit in events that one does not wish to feel complicit in (though obviously concerning subject matter of different weight).

The Value of Fictional Tragedy

The traditional video game involves straightforward heroics, and the tragic *Red Dead Redemption* is simultaneously highly experimental by video game conventions and completely traditional by storytelling conventions. However, some games reject heroics as well as sacrifice-for-a-greater-good tragedy. The game *September 12th*[22] (figure 5.16) at first appears to be a straightforward kill-the-terrorists game, but after a short while players realize that killing civilians or terrorists causes their relatives to grieve, creating more terrorists. Ian Bogost has described this as the *rhetoric of failure*,[23] a general type of design that denies the player the chance to be a hero, tragic or otherwise.

I maintain that tragedy is not inherently good or bad. In chapter 1, I quoted Steven Spielberg, who held crying players to be the benchmark for when video games would become a true storytelling art form. The truth is that we have some ambivalence about crying and unhappy endings—the sentimental tearjerker movie is often looked down upon, not celebrated. We

Figure 5.16
September 12 (Newsgaming.com 2001)

simply cannot measure the quality of an artwork by the number
of tears that are shed consuming it, and we should not accept
or reject tragedy in video games because we desire to celebrate
or denounce them as an art form. The question of tragedy is
rather interesting because it is a fundamental question about
what video games can do and what we accept as game players.

Oscar Mandel described tragedy as being of a physical or
psychological nature. I have shown here that self-destruction
(physical) and complicity (psychological) are two slightly over-
lapping ways in which games can represent tragedies in their
fictions. Complicity is arguably the more interesting of the two,
since it is unique to games, and since it extends the player's
responsibilities from the protagonist to other beings in the game

world. Complicity is also effective because it creates a new kind of mirroring between player and protagonist by letting them share an unpleasant feeling of responsibility. One of the first video games that the player could only complete by failing to protect fictional allies was the 1983 *Planetfall*,[24] where Floyd the robot sacrifices himself to save the protagonist. Prominent later examples include two 1997 games, *Final Fantasy VII*,[25] in which the romantic interest Aeris is brutally murdered, and *The Last Express*,[26] in which the heroine—in 1914—deserts the protagonist with the promise that "this war won't last longer than a month." Complicity can furthermore be integrated with both the regular experience of a sympathetic protagonist unwittingly committing unsympathetic actions and the less common experience of an unsympathetic protagonist who willingly commits unsympathetic actions. The latter is especially found in experimental live action role-playing games—games that contain an element of performance in front of other players.[27]

The awkwardness of the tragic game ending (the player being responsible for the suffering of the protagonist or for other negative events) shows that we accept regular, noninteractive tragedy in part because we lack any responsibility for the suffering. In *Othello*, we do want Desdemona to survive, but we would also be outraged at the director of the play if she did; in other words, we can feel good about feeling sorry for Desdemona[28] but we feel no responsibility for her suffering. In this sense, the fundamental difference between tragedy in games and stories is that in stories we never feel responsible for failure and suffering; in games, we do.

Still, we probably do not feel entirely responsible for tragic events in a game since they are neither real nor entirely within our control. After all, for any linear game, it is the game designer who designed the suffering and made it unavoidable[29]—in *Red Dead Redemption*, the tragic moment in the game is the one where we are relieved of agency.

Again, tragic endings are not inherently better than other types of endings, but where we paradoxically seem to enjoy witnessing unpleasant events in regular storytelling, we experience a more direct feeling of responsibility in games, making the paradox of fictional tragedy stronger, if anything, here. While no commercially successful game has offered the full *Anna Karenina* experience (playing a game in which the protagonist undergoes many painful experiences, through concrete effort managing to make the protagonist commit suicide and knowing all along that this is the goal of the game), both experimental and commercial games have offered partial versions of this. The news is not that games can present painful events (they can), but that they offer new and unique ways of doing so. The experience of complicity is a completely new type of experience that is unique to games, more personal and stronger than simply witnessing a fictional character performing the same actions. Complicity incidentally also solves the other problem presented by the straightforward game tragedy, where there was a disconnect between the player's overcoming personal flaws and deficiencies in order to complete the game, but where the protagonist had to meet a tragic end due to insoluble personal flaws. With complicity, the player shares with the protagonist the feeling of being flawed.

After the shower scene murder in Hitchcock's movie *Psycho*,[30] another well-known scene has protagonist Norman Bates attempting to hide Marion's body by stashing it in the trunk of a car, driving the car into a swamp, and watching the car sink agonizingly slowly. While we presumably find the murder repulsive, it is probably also impossible to watch this scene without identifying, however briefly, with the psychologically disturbed Bates and his quest to hide the crime. Yet our identification, and our considerations of how to best complete the crime, are entirely private: we can deny that we ever entertained such terrible thoughts. Games are more candid and direct because we have no access to such denials. The game events did not actually happen, of course; we did not endorse them by playing them; and designers have laid them out as possibilities for us. But if we were to play *Othello*, or *Anna Karenina: The Game*, we would have to admit that we thought about how to bring about the unfortunate events. All of which suggests that games are the strongest art form yet for the exploration of tragedy and responsibility. We really did consider the logistics of how to commit, or cover up, the crime. Games give us nowhere to hide.

6 The Art of·Failure

I began this short book by asking why we play games though
they contain at their core something unpleasant: the experience
of being informed that *we*, the players, are not good enough.
This paradox of failure is parallel to the paradox of tragedy and
part of the general paradox of painful art, and I argued that
these paradoxes are best understood as a conflict between two
simultaneous desires with different time frames. Our moment-
to-moment desire to avoid unpleasant experiences (for ourselves or for fictional characters) is at odds with a longer-term
aesthetic desire in which we understand failure, tragedy, and
general unpleasantness to be necessary for our experience. While
I cannot aspire to solve two thousand years of discussion of these
paradoxes, they do seem to be less mysterious once we grant
that humans can have several contradictory desires at the same
time. Moreover, where the paradox of tragedy is the domain of
philosophers, the paradox of failure is one that is constantly
discussed in our conversations about game playing: ideas of
sportsmanship tell us to control our disappointment when failing, in order to promote the general joy of the game, of partici-
pation, and of the possible positive moral implications of playing.
Not that we agree on any such explanation of the paradox of

ure, but this is a feature of games: because it is unclear what it ultimately means to fail in a game, we have the option to take game failure seriously—or to deny that our failure was important at all. Therein lies a partial freedom to fail.

The paradox of failure reappears in the psychology of failure: we are self-serving creatures inclined to evade responsibility for failure, but in order to improve our skills, we have to accept that a failure is our fault. Hence the often paradoxical guides for how to reach peak achievements: in order to win, do not play to win but to learn. Once we accept responsibility, failure also concretely pushes us to search for new strategies and learning opportunities in a game. Failure reveals strategic depth to us, and players of single-player games in particular often *need* to be pushed toward that experience. We could theoretically seek out depth and improve our skills without failing, but failure has the double function of creating in us a feeling of being flawed and forcing us to reconsider our strategies in order to escape that feeling. Interesting

Failure also feels different depending on the path to success that a game is built around. Games of skill most directly express that we are personally flawed, but failing in a game of chance encourages the less rational feeling of being on poor terms with higher powers. This does not have to be a religious sentiment but can simply be the feeling of having an unlucky day. Games of labor—of time investment—make the clearest promise that we will succeed, but at the same time they have the weakest personal implications when failing. In a longer-term perspective, different goals integrate failure into our lives and our self-image differently: a lost game with a transient goal (such as a game of tennis) is forever lost, but we are correspondingly free from having to struggle to repair our flaw since we cannot return to *that* game, only hope for revenge in a *new* game. When we begin playing a game with a completable goal, we assume the flaw of being someone who has not completed the game yet; once successful, we will always be a person who has completed the game. Compare this to improvement goals—a lifelong struggle for continually improving our own performance.

Finally, most video games represent our failures and successes by letting our performance be mirrored by a protagonist (or society, etc.) in the game's fictional world. When we are unhappy to have failed, a fictional character is also unhappy. Though the idea of a tragic game can therefore at first seem to be wholly illogical, given that the player needs to succeed by making a fictional protagonist unhappy, an undercurrent of experimental games lets the player do just that. This creates an entirely new variation of the paradox of tragedy, but the self-destruction of the protagonist remains awkward, and the better variation on game tragedy concerns our regret over complicity with events that we did not actually wish to occur. This type of tragedy is in many ways stronger than regular, nongame tragedy because we are forced to admit that we really *did* consider how to bring about the unfortunate events at the end of the game. Regular noninteractive stories shield us from responsibility in a way that games do not.

The problem of fictional game tragedy also showed that it is failure that makes us feel responsible for the events in the fictional world. Consider a game of pure labor, a game that requires only time investment to complete. In terms of structure, playing through such a game would not be too different from stepping through a movie frame by frame. Yet only a game communicates to us that *we* have not succeeded yet, and we therefore feel responsible only for positive or negative events in the fictional world when playing a game. If we play a music game such as *Guitar Hero*,[1] it is similarly only because our failure to hit the correct buttons results in an absence of music that we feel as if we are genuinely playing music when we succeed.

This provides another opportunity to return the paradox of tragedy. Control theory argues that we endure tragedy and other art forms with unpleasant content because we have control over stopping the movie, closing the book, or leaving the theater. The paradox of game tragedy shows that the argument can be reversed to a *lack-of-control* theory: perhaps we accept regular tragedy because we *do not* have control. Hamlet dies but it is not our fault (we probably do want him to die in order to give us aesthetic pleasure, but it was Shakespeare who made that call, not us). This incidentally explains why we readily indulge in historical and biographical tragedies as well as fictional ones—because nothing can be done to change the course of events, and we are free of guilt in our lack of control. Game tragedies are more direct, and harder to create, for the very same reason.

Failing Everywhere?

Games are great generators of motivation and learning, so why not use game design principles elsewhere? Educational and serious games have been explored continuously at least since the early 1970s.[2] More recently, there have been attempts to apply game design to disparate tasks such as navigating a city,[3] finding solutions to the world's energy problems,[4] rewarding consumers for brand loyalty,[5] or motivating employees.[6] Game designers have demonstrably found many ways to motivate and reward users, but, for subtle reasons, game structures can easily become counterproductive when applied to nongame contexts.

A business book makes the optimistic case that fictional employee Jennifer, demotivated by the drudgery of her call center job, will be motivated if game elements are introduced to her workplace: "The first thing Jennifer does is check on her team's progress. After the last shift, how do they rank on number of call resolutions, who in the group has 'leveled up' (achieved a new status based on performance), and who needs encouragement? All of the once-familiar call center metrics are now cast as points, ranks, and virtual currency within a large and engaging multiplayer game."[7] The argument is simple enough: clear goals and feedback are important in games. They create a general measure of our performance; they let us know how we are progressing, and they communicate when we fail and when we succeed, so let us apply them to an otherwise uninteresting work situation. Yet the example of Jennifer is illustrative in that there is little substantial difference between the "once-familiar call center metrics" and the game terminology of points, ranks, and virtual currencies. The example inadvertently reminds us that many organizations and companies are already using goals and feedback, known not as elements of "games" but as *performance measures*: the organization describes general measures of the performance of employees, collects and compares these data to performance goals, and perhaps uses the data for promotions, layoffs, and bonuses. In theory, such gamelike setups can borrow some of the strength and motivation from games: through clear goals and continuous feedback, employees can optimize their work for the good of the organization (and their own enjoyment). Furthermore, managers receive clear data about the performance of the organization. What could possibly go wrong?

Consider the experience of Sheri Zaback, a mortgage screener at the US bank Washington Mutual (WaMu) up until its collapse in 2008:

She ran applications through WaMu's computer system for approval. If she needed more information, she had to consult with a loan officer—which she described as an unpleasant experience. "They would be furious," Ms. Zaback said. "They would put it on you, that they weren't going to get paid if you stood in the way."

On one loan application in 2005, a borrower identified himself as a gardener and listed his monthly income at $12,000, Ms. Zaback recalled. She could not verify his business license, so she took the file to her boss, Mr. Parsons.

. . . a photo of the borrower's truck emblazoned with the name of his landscaping business went into the file. Approved.[8]

As can be seen, Washington Mutual had succeeded in setting up performance measures that gave a quantitative measure of the efficiency of its employees (clear goals), and at least some of them were optimizing their behavior exclusively toward the incentives that the company provided (feedback). Soon after, Washington Mutual collapsed from the losses incurred by offering loans to customers unable to pay. Seen in this light, the 2008 financial crisis was an extreme case of gamelike structures being used in the wrong way.

If clear goals and feedback can give terrible results when applied as performance goals outside games, why do they work so well in games? Part of the difficulty in creating performance goals is that they rarely measure what they are supposed to measure. For example, Soviet chandelier manufacturers were at one point rewarded based on the total weight of their output and consequently produced the heaviest chandeliers in the world.[9] As can be seen, the use of gamelike performance measures in the workplace comes across simultaneously as the latest in dynamic business management and as an antiquated bureaucratic Eastern Bloc structure. As it happens, games do not have the problem of measurement because they are ideal instances of performance goals: games have no difference between the intended output (success and failure) and the performance measure (success and failure). The value system that the goal of a game creates is not an artificial measure of the value of the player's performance; the goal is what *creates* the value in the first place by assigning values to the possible outcomes of a game.

The good news is that game design methods can be of great help here: games push players to optimize for performance at the expense of other considerations, and players tend to pick the path of least resistance. A game designer must always ask whether the optimal path for the player is the right one; does the game push the player toward taking the most interesting and enjoyable path, or toward taking an uninteresting path? It is the responsibility of the designer to make sure that the optimal path is the right one. A similar question must be asked when setting performance goals: what will happen if users actually follow the optimal path of achieving the performance goal? Viewing performance measures through the lens of game design should therefore help us avoid the worst potential problems in using game structures for nongame purposes.

Return to the Paradox of Failure

Much of this book has been concerned with the question of whether game failure is painful. We have a range of ways to dismiss the importance of failure in games. Such is the fundamental duality of failure in games: games can by their very definition be played without any tangible consequences,[10] but they give us a license to care about playing even when it has no obvious benefits. Then, they give us the additional license to deny that we dislike failing at the same task. This license is central to why games work so well, even though they make us fail. Some psychologists believe that adding tangible benefits can make us *less*, not more, interested in an activity that we find motivating.[11] Games are not motivating *despite* the fact that they give us no tangible rewards but *because* they give us no tangible rewards, and hence no tangible punishment for failure.

Outside games, when failure has certain consequences, we are more likely to stick to safe strategies, minimizing the learning and personal improvement that could otherwise have taken place. Games, then, are the best possible versions of challenges, learning, and failure *as such*.[12] It follows that when game structures are applied directly to activities with tangible consequences, the plausible deniability of failure may disappear: if our performance is seen as a genuine measure of who we are, then we can no longer claim that our failure was unimportant. The freedom found in regular games can only be preserved if we are given room to experiment and the freedom to fail, at least temporarily, such that a single poor performance will not be used against us. We will retain deniability to ourselves—and toward teachers, colleagues, and supervisors.[13]

Failure forces us to reconsider what we are doing, to learn. Failure connects us personally to the events in the game; it proves that *we matter*, that the world does not simply continue regardless of our actions. Still, the question of game failure has a recursive quality. For example, Apter's reversal theory seems to explain why we would seek out failure and danger in games, yet players often seek boring strategies in order to avoid failure once a game has begun. For every explanation of how different, or similar, game failure is to regular out-of-game failure, an exception can be found, showing that the picture is not complete. A tempting solution could be to say that the weight of game failure is purely subjective, that it means exactly what we want it to mean, but that overstates how much control we have. Perhaps we would *like* game failure to be what we make it, but that is not within our power. The paradox of failure, like the paradox of tragedy, and the general paradox of painful of art, does not describe a universal agreement to seek out emotions that we would otherwise abhor. There are tragedies that we are not willing to witness, horror movies that we are unwilling to watch.[14] And there are games that we refuse to play with siblings, rivals, bosses, ourselves, because we are too afraid of failure.

Perhaps we would like to behave nobly, according to a higher code, as perfect sportspeople. Yet the history of games is scattered with examples of even top players who fail that test. After winning the world chess championship over Spassky in 1972, Bobby Fischer began making a series of increasingly unreasonable demands for future matches, eventually forfeiting his title. According to former world champion Garry Kasparov, the always meticulously prepared Fischer was so afraid of failing that he preferred not to play.[15]

In probably the most famous quotation in soccer history, Bill Shankly once declared, "Some people believe football is a matter of life and death. I'm very disappointed with that attitude. I can assure you it is much, much more important than that." I sympathize. In a way, I would like to excise from the English language the phrase "just a game," because it pretends something that is not true, that failure is neutral as long as it happens in a game. Ironically, this would be the wrong thing to do, as it is exactly this faulty claim that gives us a chance to shrug our shoulders while fuming inside. It is a pretense that allows us the freedom to count our successes and silently downplay our failures, even to ourselves. We must accept that this shiny surface of harmlessness creates a space where we can struggle with our failures and flaws. This illusive space of games is to be protected, but it must always come with an additional license for us to be just a little angry, and more than a little frustrated, when we fail. That—not balance, but strange arrangement—is games, the *art* of failure.

Notes

1 Introduction

1. Japan Studios, *Patapon*.

2. Q Entertainment, *Meteos*.

3. The *Meteos* game mode was Star Strip, Branch route, default difficulty, to be precise.

4. Przybylski, Rigby, and Ryan, "A Motivational Model of Video Game Engagement."

5. Since Aristotle only refers to catharsis once in the *Poetics*, there is a long history of conflicting interpretations of the term. See, for example, Golden, "Catharsis," or Keesey, "On Some Recent Interpretations of Catharsis."

6. Csikszentmihalyi, *Flow*.

7. Veteran game designer Noah Falstein has refined flow theory to say that game difficulty should vary in waves—sometimes the game should be a little easy, sometimes a little hard. (Falstein, "Understanding Fun—The Theory of Natural Funativity"). Such irregular increases in difficulty make the player more likely to experience both failures and successes.

8. Juul, *A Casual Revolution*.

9. Valve, *Portal 2*.

10. I am reluctantly using the term *sportsmanship* rather than the gender-neutral *sportspersonship* as the latter remains an awkward construction.

11. Arnold, "Three Approaches toward an Understanding of Sportsmanship."

12. Banks, *Liam Wins the Game, Sometimes*.

13. Ibid., 22.

14. McGonigal, *Reality Is Broken*.

15. Zichermann and Linder, *Game-Based Marketing*.

16. Teuber, *The Settlers of Catan*.

17. While designers continue to be asked to create interfaces that minimize user errors, there is also a movement toward considering the general user experience as more than just optimization for task efficiency. See, for example, Norman, *Emotional Design*.

18. Monty Python, "Dirty Hungarian Phrasebook."

19. Juul and Norton, "Easy to Use and Incredibly Difficult."

20. Bateson, "A Theory of Play and Fantasy."

21. Salen and Zimmerman, *Rules of Play*.

22. Juul, "The Magic Circle and the Puzzle Piece."

23. Lazzaro, "Why We Play," 686.

24. Namco, *Pac-Man*.

25. Rockstar San Diego & Rockstar North, *Red Dead Redemption*.

26. Playdead Studios, *Limbo*.

27. Nintendo EAD, *Super Mario Bros*.

28. New York Post, "This Sport Is Stupid Anyway."

29. Försterling, *Attribution*, 47.

30. Franklin, *The Morals of Chess*.

31. Hudson, *Chariots of Fire*.

32. Schiller, *On the Aesthetic Education of Man*, 80.

33. Huizinga, *Homo Ludens*, 13.

34. Plato, *The Republic*, 348.

35. For example, Heidegger writes about Van Gogh's painting *A Pair of Worn Shoes*, "Van Gogh's painting is the disclosure of what the equipment, the pair of peasant shoes, *is* in truth" (Heidegger, "The Origin of the Work of Art," 380).

36. Juul, *Half-Real*, chap. 2.

37. In the late 1950s, Roger Caillois defined games as "unproductive" (Caillois, *Man, Play and Games*, 10). The problem is that games can produce all kinds of material gain—look no further than the gambling industry. In *Half-Real*, I argued rather that games are characterized by the fact that their tangible consequences are negotiable—every game can be played with or without tangible consequences (such as money being exchanged).

38. Juul, *Half-Real*.

39. Mandel, *A Definition of Tragedy*, 88.

40. Nietzsche, *The Birth of Tragedy and Other Writings*, 113.

41. Sutton-Smith, "Play as a Parody of Emotional Vulnerability," 3.

42. BioWare, *Mass Effect 2*.

43. Ryan, "Beyond Myth and Metaphor."

44. Juul, *Half-Real*, 161.

45. Rockstar San Diego & Rockstar North, *Red Dead Redemption*.

46. Breznican, "Spielberg, Zemeckis Say Video Games, Films Could Merge."

47. Blizzard Entertainment, *World of Warcraft*.

48. Square, *Final Fantasy VII*.

49. Zynga, *FarmVille*.

2 The Paradox of Failure and the Paradox of Tragedy

1. Namco, *Pac-Man*.

2. Via the *Ludologist* blog, www.jesperjuul.net/ludologist.

3. Juul, "Fear of Failing?

4. The result for all three categories of player performance combined was close to statistical significance ($p = 0.06$).

5. Smuts, "The Paradox of Painful Art," 60.

6. Smuts, "The Paradox of Painful Art."

7. Levinson, "Emotion in Response to Art."

8. Walton, *Mimesis as Make-Believe*.

9. Coleridge, *Biographia Literaria*, ch. xiv.

10. Yanal, *Paradoxes of Emotion and Fiction*, 102.

11. Eaton, "A Strange Kind of Sadness."

12. Neill, "On a Paradox of the Heart," 54.

13. Aristotle, *Poetics*, 47–49.

14. Golden, "Catharsis."

15. Carroll, *The Philosophy of Horror*, 186.

16. Feagin, "The Pleasures of Tragedy."

17. Neill, "On a Paradox of the Heart," 60.

18. Ibid.

19. Shakespeare, *Othello*.

20. While Currie is probably correct in pointing out that a modern audience would be outraged if Desdemona survived in *Othello*, Shakespeare's *King Lear* was for a long time mostly performed in Nahum Tate's happy-ending variation *The History of King Lear*. (This was pointed out to me by Clara Fernández-Vara.)

21. Currie, "Tragedy."

22. Grodal, *Moving Pictures*.

23. Frome, "Representation, Reality, and Emotions across Media."

24. Banks, *Liam Wins the Game, Sometimes*.

25. The Olympic Museum, "The Olympic Symbols."

3 The Feeling of Failure

1. Sega Wow Inc., *Super Real Tennis*.

2. The complete "basic operation methods" are as follows: "To move around: Towards the net, use Up key. Towards the baseline, use Down key. To the right, use Right key, and to the left, use Left key. To swing the racket, center press or use key 5."

3. Lardon, *Finding Your Zone*, 96.

4. Max, "The Prince's Gambit."

5. Lardon, *Finding Your Zone*, 106–110.

6. Abramson, Seligman, and Teasdale, "Learned Helplessness in Humans."

7. Nintendo SDD, *Brain Age*.

8. Juul, "Fear of Failing?"

9. The result for answers "The game was too hard" and "I made a mistake" was statistically significant ($p < 0.016$).

10. Singapore-MIT GAMBIT Game Lab, *Pierre*.

11. Yates, "Carcassonne Strategy."

12. Järvinen, "Games without Frontiers."

13. Zillmann, "Mood Management through Communication Choices."

14. Järvinen, "Games without Frontiers," 111.

15. Butler and McManus, *Psychology*, 112.

16. Diener and Dweck, "An Analysis of Learned Helplessness."

17. Squire, "Changing the Game."

18. Zuckerman, "Attribution of Success and Failure Revisited," 246.

19. Weiner, "'Spontaneous' Causal Thinking."

20. This model has some similarities to Arsenault and Perron's "magic cycle" (Arsenault and Perron, "In the Frame of the Magic Cycle").

21. Jaffe, "Aaaaaaaaannnnnnnnnddddddd Scene!"

22. Firaxis Games, *Civilization III*.

23. Johnson, "Water Finds a Crack."

24. Kerr and Apter, *Adult Play*, 17.

25. Smith, "Plans and Purposes," 217–227.

26. Rockstar North, *Grand Theft Auto IV*.

27. Deppe and Harackiewicz, "Self-Handicapping and Intrinsic Motivation."

28. EA Black Box, *Skate 2*.

29. Juul, "Without a Goal."

30. Maxis, *SimCity*.

4   Designing Failure

1. Pickford, "The First 30 Seconds."

2. "Super Meat Boy brings the old school difficulty of classic retro titles we all know and love and stream lines them down to the essential no bull straight forward twitch reflex platforming" (Xbox.com, "Super Meat Boy").

3. Saltsman, "Game Design Accessibility Matters."

4. Naughty Dog, *Uncharted 2*.

5. messhoff, *Flywrench*.

6. Team Meat, *Super Meat Boy*.

7. With the complication that many arcade games offered extra lives for scores above certain thresholds, and adventure games offered infinite retries.

8. Orland, "The Slow Death of the Game Over."

9. In economic game theory, Neumann and Morgenstern distinguished between (physical) skill and (mental) strategy, but I will treat skill and strategy as one here (Neumann and Morgenstern, *Theory of Games and Economic Behavior*).

10. PopCap Games, *Bejeweled 2 Deluxe*.

11. Blizzard Entertainment, *World of Warcraft*.

12. Clark, "A Fantasy of Labor."

13. Juul, *Half-Real*, chap. 3.

14. Gee, *What Video Games Have to Teach Us about Learning and Literacy*.

15. Kent, *The First Quarter*, 72–73.

16. Supreme Court of Denmark case 191/2008, *Pokerspillet Texas Hold'em i turneringsform anset for hasard*.

17. As a bare minimum, judges considering whether poker is a game of skill should be required to play a few hands against an expert player to see how well they would fare.

18. Zynga, *FarmVille*.

19. Kushner, "Games."

20. Cliffe, *Counter-Strike*.

21. Lazzaro, "The 4 Most Important Emotions of Social Games."

22. Backburner Games, *Statbuilder*.

23. Linderoth, "Why Gamers Don't Learn More."

24. id Software, *Doom*.

25. Sirlin, "Soapbox."

26. I am interpreting the three kinds of fairness according to the framework of Almås et al., "Fairness and the Development of Inequality Acceptance."

27. Scott Rettberg argues in particular that *World of Warcraft* embodies the Protestant work ethic as expressed by capitalism (Rettberg, "Corporate Ideology in World of Warcraft"). One interesting note is that many Asian video games are heavily labor-based, and hence the connection is surely more complicated.

28. Almås et al., "Fairness and the Development of Inequality Acceptance."

29. At the time of writing, the economic model of most free-to-play games is focused on selling the user both special decorative items and shortcuts that allow the user to progress faster in the game.

30. Bogost, *Persuasive Games*.

31. Flanagan et al., "A Method For Discovering Values in Digital Games."

32. Parker Brothers, *Monopoly*.

33. Magie, *The Landlord's Game*.

34. Philips, "Game-Board."

35. The Olympic Museum, "The Olympic Symbols."

36. BioWare, *Mass Effect 2*.

37. It should be said the *Mass Effect* series is unusual in that some failures will kill off the player's fictional teammates, but not prevent the player from completing the game. Another recent example of this is Quantic Dream's *Heavy Rain*, which can be completed despite the death of central characters.

38. Dimps/Capcom, *Super Street Fighter IV*.

39. In practice, many older arcade games have bugs, *kill screens*, that make them unplayable past certain levels, and hence they do not

actually support the infinite personal improvement that they seem to encourage.

40. Backbone Entertainment, *Bomberman Live*.

41. Ubisoft Montreal, *Far Cry 2*.

42. Abraham, "Permanent Death, Episode 1."

43. Taito, *Space Invaders*.

5 Fictional Failure

1. Murray, *Hamlet on the Holodeck*.

2. For a longer discussion of the different goals that a player can be aligned with, see Montola, "The Invisible Rules of Role-Playing."

3. This disconnect was identified to me by Jonathan Frome.

4. Psychology distinguishes between emotional *valence* and *arousal*: valence refers to the positive or negative action potential of an emotion (whether it causes us to seek or avoid what caused it) and arousal (how exciting it is). Horror or tragedies have high arousal (our hearts beat faster, we perspire) at the same time that they have negative valence (unpleasant subject matter). (Lang et al., "Looking at Pictures").

5. Juul, *A Casual Revolution*, chap. 2.

6. A more detailed discussion of perspective in video games can be found in Thon, "Perspective in Contemporary Computer Games."

7. Ryan, "Beyond Myth and Metaphor."

8. Juul, *Half-Real*, 161.

9. The *Othello* example is also hard to imagine as a game since the central plot is based on the Iago's deception of Othello. Should the player controlling Othello be aware of Iago's deception or not?

10. Murray, *Hamlet on the Holodeck*, 25.

11. Tolstoy, *Anna Karenina*.

12. Dang and Li, *The Suicide Game*.

13. Criterion Games, *Burnout Paradise*.

14. Secret Exit, *Stair Dismount*.

15. Team Meat, *Super Meat Boy*.

16. SCE Studios Santa Monica, *God of War*.

17. Rockstar San Diego & Rockstar North, *Red Dead Redemption*.

18. Kazicun, "Poll: Red Dead Redemption Ending Closing Comments (*SPOILERS*)."

19. Brathwaite, *Train*.

20. Elgot, "Monopoly, Cluedo, Holocaust the Board Game?"

21. Team Ico, *Shadow of the Colossus*.

22. Newsgaming.com, *September 12th*.

23. Bogost, *Persuasive Games*, 2.

24. Infocom, *Planetfall*.

25. Square, *Final Fantasy VII*.

26. Smoking Car Productions, *The Last Express*.

27. Montola, "The Positive Negative Experience in Extreme Role-Playing."

28. Currie, "Tragedy."

29. Miguel Sicart has explored the argument that games offer a "banality of evil" in that players can always deny responsibility by claiming that they acted under orders (Sicart, "The Banality of Simulated Evil").

30. Hitchcock, *Psycho*.

6 The Art of Failure

1. Harmonix Music Systems, Inc., *Guitar Hero*.

2. Clark C. Abt, *Serious Games*.

3. The Foursquare service offers badges and rewards for "checking in" to different locations such as coffee shops.

4. Independent Lens, *World without Oil*.

5. Zichermann and Linder, *Game-Based Marketing*.

6. Reeves and Read, *Total Engagement*.

7. Ibid., 2.

8. Goodman and Morgenson, "Saying Yes, WaMu Built Empire on Shaky Loans."

9. Courty and Marschke, "Dynamics of Performance-Measurement Systems."

10. Juul, *Half-Real*, chap. 2.

11. Deci, "Effects of Externally Mediated Rewards on Intrinsic Motivation."

12. Games are quite similar to more work-like activities in that they require considerable effort, but because games can be about any possible subject matter, designers are free to construct them without regards to the constraints of regular activities. This is why it is harder to design educational games than noneducational games—designers of educational games will always lack some of the freedom that other game designers enjoy.

13. In many cases, it may be impossible to retain playfulness in the face of performance measures: there seems to be a law that regardless of how weak and uncertain a data point is, someone is always going to take it to be an objective measure of truth.

14. McCauley, "When Screen Violence Is Not Attractive."

15. Kasparov, "The Bobby Fischer Defense."

Bibliography

Abraham, Ben. "Permanent Death, Episode 1: An Inauspicious Beginning." *SLRC—Subterranean Loner Rendered Comatose*, June 24, 2009. http://drgamelove.blogspot.com/2009/06/permanent-death-episode-1 -inasupicious.html.

Abramson, L. Y., M. E. Seligman, and J. D. Teasdale. "Learned Helplessness in Humans: Critique and Reformulation." *Journal of Abnormal Psychology* 87 (1) (February 1978): 49–74.

Abt, Clark C. *Serious Games*. New York: Viking Press, 1970.

Almås, Ingvild, Alexander W. Cappelen, Erik Ø. Sørensen, and Bertil Tungodden. "Fairness and the Development of Inequality Acceptance." *Science* 328 (5982) (May 28, 2010): 1176–1178.

Aristotle. *Poetics*. Trans. Stephen Halliwell. Cambridge, MA: Harvard University Press, 1999.

Arnold, Peter J. "Three Approaches toward an Understanding of Sportsmanship." In *Philosophic Inquiry in Sport*, 2nd ed., ed. William J. Morgan and Klaus V. Meier, 161–167. Champaign, IL: Human Kinetics, 1995.

Arsenault, Dominic, and Bernard Perron. "In the Frame of the Magic Cycle: The Circle(s) of Gameplay." In *The Video Game Theory Reader 2*, ed. Bernard Perron and Mark J. P. Wolf, 109–132. New York: Taylor & Francis, 2009.

Backbone Entertainment. *Bomberman Live*. Hudson Soft (Xbox 360), 2007.

Backburner Games. *Statbuilder*. Kongregate (Flash), 2008. http://www
.kongregate.com/games/backburner/statbuilder-classic.

Banks, Jane Whelen. *Liam Wins the Game, Sometimes: A Story about Losing
with Grace (Liam Says)*. London: Jessica Kingsley Publishers, 2009.

Bateson, Gregory. "A Theory of Play and Fantasy." In *Steps to an Ecology
of Mind*, 177–193. Chicago, IL: University of Chicago Press, 2000.

BioWare. *Mass Effect 2*. Electronic Arts (Xbox 360), 2010.

Blizzard Entertainment. *World of Warcraft*. Blizzard Entertainment
(Windows), 2004.

Bogost, Ian. *Persuasive Games: The Expressive Power of Videogames*. Cambridge, MA: MIT Press, 2007.

Brathwaite, Brenda. *Train*. Board game, 2009.

Breznican, Anthony. "Spielberg, Zemeckis Say Video Games, Films
Could Merge." *USA Today*, September 16, 2004. http://www.usatoday.com/
tech/products/games/2004-09-16-game-movie-meld_x.htm.

Butler, Gillian, and Freda McManus. *Psychology: A Very Short Introduction*.
Oxford: Oxford University Press, 1998.

Caillois, Roger. *Man, Play and Games*. Urbana: University of Illinois Press,
2001.

Carroll, Nöel. *The Philosophy of Horror*. New York: Routledge, 1990.

Clark, Naomi. "A Fantasy of Labor." Paper presented at Games for Change,
New York, May 26, 2010.

Cliffe, Minh "Gooseman" Le Jess. *Counter-Strike*. Vivendi Universal
(Windows), 1999.

Coleridge, Samuel Taylor. *Biographia Literaria*, 1817. http://www
.gutenberg.org/ebooks/6081.

Courty, Pascal, and Gerald Marschke. "Dynamics of Performance-
Measurement Systems." *Oxford Review of Economic Policy* 19 (2) (2003):
268–284.

Criterion Games. *Burnout Paradise: Ultimate Box*. Electronic Arts. (Windows), 2009.

Csikszentmihalyi, Mihaly. *Flow: The Psychology of Optimal Experience*. New York: Harper & Row, 1990.

Currie, Gregory. "Tragedy." *Analysis* 70 (4) (September 2010): 632–638.

Dang, Albert, and Kan Yang Li. *The Suicide Game*, 2006. http://www .jesperjuul.net/text/suicidegame/.

Deci, Edward L. "Effects of Externally Mediated Rewards on Intrinsic Motivation." *Journal of Personality and Social Psychology* 18 (1) (1971): 105–115.

Deppe, R. K., and J. M. Harackiewicz. "Self-Handicapping and Intrinsic Motivation: Buffering Intrinsic Motivation from the Threat of Failure." *Journal of Personality and Social Psychology* 70 (4) (April 1996): 868–876.

Diener, C. I., and C. S. Dweck. "An Analysis of Learned Helplessness: Continuous Changes in Performance, Strategy, and Achievement Cognitions Following Failure." *Journal of Personality and Social Psychology* 36 (5) (1978): 451–462.

Dimps/Capcom. *Super Street Fighter IV*. Capcom (Xbox 360), 2010.

EA Black Box. *Skate 2*. Electronic Arts (Xbox 360), 2009.

Eaton, Marcia M. "A Strange Kind of Sadness." *Journal of Aesthetics and Art Criticism* 41 (1) (October 1, 1982): 51–63.

Elgot, Jessica. "Monopoly, Cluedo, Holocaust the Board Game?" *The Jewish Chronicle*, October 28, 2010. http://www.thejc.com/news/world -news/40318/monopoly-cluedo-holocaust-board-game.

Falstein, Noah. "Understanding Fun—The Theory of Natural Funativity." In *Introduction to Game Development*, 1st ed., ed. Steve Rabin, 71–98. Rockland, MA: Charles River Media, 2005.

Feagin, Susan L. "The Pleasures of Tragedy." *American Philosophical Quarterly* 20 (1) (January 1983): 95–104.

Firaxis Games. *Civilization III*. Infogrames (Windows), 2001.

Flanagan, Mary, Helen Nissenbaum, Jonathan Belman, and Jim Diamond. "A Method for Discovering Values in Digital Games." In *Situated Play: Proceedings of the Third International Conference of the Digital Games Research Association (DiGRA)*. Tokyo, 2007. http://www.digra.org/dl/db/07311.46300.pdf.

Försterling, Friedrich. *Attribution: An Introduction to Theories, Research and Applications*. London: Psychology Press, 2001.

Franklin, Benjamin. *The Morals of Chess*. Philadelphia, PA: Columbian Magazine, 1786.

Frome, Jonathan. "Representation, Reality, and Emotions across Media." *Film Studies: An International Review* 8/9 (Spring 2006): 12–25. http://www.jonathanfrome.net/papers/representation-reality.html.

Gamehelper. *Skate 2—Best of Reel*, 2009. http://www.youtube.com/watch?v=oWkh0AsKBJc.

Gee, James Paul. *What Video Games Have to Teach Us about Learning and Literacy*. New York: Palgrave Macmillan, 2003.

Golden, Leon. "Catharsis." *Transactions and Proceedings of the American Philological Association* 93 (1962): 51–60.

Goodman, Peter S., and Gretchen Morgenson. "Saying Yes, WaMu Built Empire on Shaky Loans." *The New York Times*, December 28, 2008, sec. Business. http://www.nytimes.com/2008/12/28/business/28wamu.html.

Grodal, Torben. *Moving Pictures: A New Theory of Film Genres, Feelings, and Cognition*. New York: Oxford University Press, 1999.

Harmonix Music Systems, Inc. *Guitar Hero*. RedOctane Inc. (PlayStation 2), 2005.

Heidegger, Martin. "The Origin of the Work of Art." In *Twentieth Century Theories of Art*, ed. James Matheson Thompson. Montreal, QC: McGill-Queen's Press, 1990.

Hitchcock, Alfred. *Psycho*. Paramount Pictures, 1960.

Hudson, Hugh. *Chariots of Fire*. 20th Century Fox, 1981.

Huizinga, Johan. *Homo Ludens*. Boston, MA: Beacon Press, 1950.

id Software. *Doom*. GT Interactive (DOS), 1993.

Independent Lens. *World without Oil*. Alternate Reality Game, 2007. http://www.worldwithoutoil.org/.

Infocom. *Planetfall*. Infocom (DOS), 1983.

Jaffe, David. "Aaaaaaaaannnnnnnnddddddd Scene!" *Jaffe's Game Design*, November 25, 2007. http://criminalcrackdown.blogspot.com/2007_11 _25_archive.html.

Japan Studios. *Patapon*. Sony Computer Entertainment (PSP), 2008.

Järvinen, Aki. "Games without Frontiers." PhD dissertation, Tampere University, Finland, 2008. http://acta.uta.fi/english/teos.php?id=11046.

Johnson, Soren. "Water Finds a Crack: How Player Optimization Can Kill a Game's Design." *Game Developer Magazine*, March 2011, 32–33.

Juul, Jesper. *A Casual Revolution: Reinventing Video Games and Their Players*. Cambridge, MA: MIT Press, 2009.

Juul, Jesper. "Fear of Failing? The Many Meanings of Difficulty in Video Games." In *The Video Game Theory Reader 2*, ed. Bernard Perron and Mark J. P. Wolf, 237–252. New York: Routledge, 2008.

Juul, Jesper. *Half-Real: Video Games between Real Rules and Fictional Worlds*. Cambridge, MA: MIT Press, 2005.

Juul, Jesper. "The Magic Circle and the Puzzle Piece." In *Conference Proceedings of The Philosophy of Computer Games 2008*, ed. Stephan Günzel and Dieter Mersch, 56–69. Potsdam: Potsdam University Press, 2008.

Juul, Jesper. "Without a Goal: On Open and Expressive Games." In *Videogame, Player, Text*, ed. Barry Atkins and Tanya Krzywinska, 191–203. Manchester: Manchester University Press, 2007. http://www.jesperjuul .net/text/withoutagoal.

Juul, Jesper, and Marleigh Norton. "Easy to Use and Incredibly Difficult: on the Mythical Border between Interface and Gameplay." In *Proceedings of the 4th International Conference on Foundations of Digital Games*, 107–112. Orlando, FL: ACM, 2009. http://portal.acm.org/citation.cfm?id=1536539.

Kasparov, Garry. "The Bobby Fischer Defense." *New York Review of Books*, March 2011, 10. http://www.nybooks.com/articles/archives/2011/mar/10/bobby-fischer-defense/.

Kazicun. "Poll: Red Dead Redemption Ending Closing Comments (*SPOILERS*)," May 23, 2010. http://www.escapistmagazine.com/forums/read/9.196713-Poll-Red-Dead-Redemption-Ending-Closing-Comments-SPOILERS.

Keesey, Donald. "On Some Recent Interpretations of Catharsis." *Classical World* 72 (4) (January 1978): 193–205.

Kent, Steven L. *The First Quarter: A 25-Year History of Video Games*. Bothell, WA: BWD Press, 2000.

Kerr, J. H., and J. Michael Apter. *Adult Play: A Reversal Theory Approach*. Amsterdam: Swets & Zeitlinger, 1991.

Kushner, David. "Games: Why Zynga's Success Makes Game Designers Gloomy." *Wired*, September 27, 2010. http://www.wired.com/magazine/2010/09/pl_games_zynga/.

Lang, Peter J., Mark K. Greenwald, Margaret M. Bradley, and Alfons O. Hamm. "Looking at Pictures: Affective, Facial, Visceral, and Behavioral Reactions." *Psychophysiology* 30 (3) (1993): 261–273.

Lardon, Michael. *Finding Your Zone: Ten Core Lessons for Achieving Peak Performance in Sports and Life*. New York: Perigee Trade, 2008.

Lazzaro, Nicole. "The 4 Most Important Emotions of Social Games." Paper presented at Game Developers Conference, San Francisco, CA, March 9–13, 2010.

Lazzaro, Nicole. "Why We Play: Affect and the Fun of Games." In *The Human-Computer Interaction Handbook: Fundamentals, Evolving Eechnologies, and Emerging Applications*, 2nd ed., ed. Andrew Sears and Julie A. Jacko, 679–700. New York: Lawrence Erlbaum Associates, 2008.

Level-5. *Professor Layton and the Curious Village*. Nintendo (DS), 2008.

Levinson, Jerrold. "Emotion in Response to Art." In *Emotion and the Arts*, ed. Mette Hjort and Sue Laver, 20–34. Oxford: Oxford University Press, 1997.

Linderoth, Jonas. "Why Gamers Don't Learn More: An Ecological Approach to Games as Learning Environments." In *Nordic DiGRA 2010 Proceedings*. Stockholm, 2010. http://www.digra.org/dl/db/10343.51199.pdf.

Magie, Elisabeth. *The Landlord's Game*. Board game, 1910.

Mandel, Oscar. *A Definition of Tragedy*. New York: New York University Press, 1961.

Max, D. T. "The Prince's Gambit." *New Yorker*, March 21, 2011, 21. http://www.newyorker.com/reporting/2011/03/21/110321fa_fact_max.

Maxis. *SimCity*. Maxis (DOS), 1989.

McCauley, Clark. "When Screen Violence Is Not Attractive." In *Why We Watch: The Attractions of Violent Entertainment*, ed. Jeffrey Goldstein, 144–162. New York: Oxford University Press, 1998.

McGonigal, Jane. *Reality Is Broken: Why Games Make Us Better and How They Can Change the World*. New York: Penguin Press, 2011.

messhoff. *Flywrench*. messhoff (Windows), 2007. http://messhof.com/flywrench/.

Montola, Markus. "The Invisible Rules of Role-Playing: The Social Framework of Role-Playing Process." *International Journal of Role-Playing* 1 (1) (2008): 22–36.

Montola, Markus. "The Positive Negative Experience in Extreme Role-Playing." In *Nordic DiGRA 2010 Proceedings*. Stockholm, 2010. http://www.digra.org/dl/db/10343.56524.pdf

Monty Python. "Dirty Hungarian Phrasebook." *Monty Python's Flying Circus*, episode 25." BBC, 1970.

Murray, Janet H. *Hamlet on the Holodeck: The Future of Narrative in Cyberspace*. Cambridge, MA: MIT Press, 1998.

Namco. *Pac-Man*. Namco (Arcade), 1980.

Naughty Dog. *Uncharted 2: Among Thieves*. Sony Computer Entertainment (PlayStation 3), 2009.

Neill, Alex. "On a Paradox of the Heart." *Philosophical Studies: An International Journal for Philosophy in the Analytic Tradition* 65 (1/2) (February 1992): 53–65.

Neumann, John Von, and Oskar Morgenstern. *Theory of Games and Economic Behavior*. Princeton, NJ: Princeton University Press, 1944.

Newsgaming.com. *September 12th*. Newsgaming.com (Flash), 2001.

New York Post. "This Sport Is Stupid Anyway." *New York Post*, June 27, 2010.

Nietzsche, F. W. *The Birth of Tragedy and Other Writings*, ed. R. Geuss and R. Speirs. Cambridge, UK: Cambridge University Press, 1999.

Nintendo EAD. *Super Mario Bros*. Nintendo (NES), 1985.

Nintendo SDD. *Brain Age*. Nintendo (DS), 2006.

Norman, Donald A. *Emotional Design: Why We Love (or Hate) Everyday Things*. New York: Basic Books, 2005.

The Olympic Museum. "The Olympic Symbols," 2007. http://multimedia.olympic.org/pdf/en_report_1303.pdf.

Orland, Kyle. "The Slow Death of the Game Over." *The Escapist*, June 5, 2007. http://www.escapistmagazine.com/articles/view/issues/issue_100/556-The-Slow-Death-of-the-Game-Over.

Parker Brothers. *Monopoly*. Parker Brothers (board game), 1936.

Philips, Elisabeth Magie. "Game-Board." US Patent no. 1,509,312. Washington, DC, September 23, 1924.

Pickford, Ste. "The First 30 Seconds." *Ste Pickford's Blog*, December 3, 2007. http://www.zee-3.com/pickfordbros/blog/view.php?post=368.

Plato. *The Republic*, ed. Desmond Lee and Sir Henry Desmond Pritchard Lee. London: Penguin Classics, 2003.

Playdead Studios. *Limbo*. Microsoft Game Studios (Xbox 360), 2010.

PopCap Games. *Bejeweled 2 Deluxe*. PopCap Games (Windows), 2004.

Przybylski, Andrew K., C. Scott Rigby, and Richard M. Ryan. "A Motivational Model of Video Game Engagement." *Review of General Psychology* 14 (2) (2010): 154–166.

Q Entertainment. *Meteos*. Ubisoft (DS), 2005.

Quantic Dream. *Heavy Rain*. Sony Computer Entertainment (PlayStation 3), 2010.

Reeves, Byron, and J. Leighton Read. *Total Engagement: Using Games and Virtual Worlds to Change the Way People Work and Businesses Compete*. Boston, MA: Harvard Business School Press, 2009.

Rettberg, Scott. "Corporate Ideology in World of Warcraft." In *Digital Culture, Play, and Identity: A World of Warcraft Reader*, ed. Hilde Corneliussen and Jill Walker Rettberg, 19–38. Cambridge, MA: MIT Press, 2008.

Rockstar North. *Grand Theft Auto IV*. Rockstar Games (Xbox 360), 2008.

Rockstar San Diego & Rockstar North. *Red Dead Redemption*. Rockstar Games (Xbox 360), 2010.

Ryan, Marie-Laure. "Beyond Myth and Metaphor: The Case of Narrative in Digital Media." *Game Studies* 1, no. 1 (2001). http://www.gamestudies.org/0101/ryan/.

Salen, Katie, and Eric Zimmerman. *Rules of Play: Game Design Fundamentals*. Cambridge, MA: MIT Press, 2004.

Saltsman, Adam. "Game Design Accessibility Matters." *Gamasutra*, January 6, 2010. http://www.gamasutra.com/view/news/26386/Analysis_Game_Design_Accessibility_Matters.php.

SCE Studios Santa Monica. *God of War*. Sony Computer Entertainment (PlayStation 2), 2005.

Schiller, Friedrich. *On the Aesthetic Education of Man*, ed. Reginald Snell. N. Chelmsford, MA: Courier Dover Publications, 2004.

Secret Exit. *Stair Dismount*. Secret Exit (iPhone), 2009. http://www.stairdismount.com/.

Sega. *Super Monkey Ball Deluxe*. Sega (Xbox), 2005.

Sega Wow Inc. *Super Real Tennis*. Sega Corporation (Mobile), 2003.

Shakespeare, William. *Othello*. Philadelphia, PA: J. B. Lippincott Company, [1603] 1886.

Sicart, Miguel. "The Banality of Simulated Evil: Designing Ethical Gameplay." *Ethics and Information Technology* 11 (3) (2009): 191–202.

Singapore-MIT GAMBIT Game Lab. *Pierre: Insanity Inspired*. Gambit (Flash), 2009. http://gambit.mit.edu/loadgame/pierre.php.

Sirlin, David. "Soapbox: World of Warcraft Teaches the Wrong Things." *Gamasutra*, February 22, 2006. http://www.gamasutra.com/view/feature/2567/soapbox_world_of_warcraft_teaches_.php.

Smith, Jonas Heide. "Plans and Purposes: How Video Games Shape Player Behavior." PhD dissertation, IT University of Copenhagen, 2006. http://jonassmith.dk/weblog/wp-content/dissertation1-0.pdf

Smoking Car Productions. *The Last Express*. Brøderbund (Windows), 1997.

Smuts, Aaron. "The Paradox of Painful Art." *Journal of Aesthetic Education* 41 (3) (2007): 59–76.

Square. *Final Fantasy VII*. Square (PlayStation), 1997.

Squire, K. "Changing the Game: What Happens When Video Games Enter the Classroom." *Innovate: Journal of Online Education* 1 (6) (2005). http://www.innovateonline.info/index.php?view=article&id=82.

Supreme Court of Denmark case 191/2008. *Pokerspillet Texas Hold'em i turneringsform anset for hasard*, 2009.

Sutton-Smith, Brian. "Play as a Parody of Emotional Vulnerability." In *Play and Educational Theory and Practice*, ed. Donald E. Lythe, 3–17. Westport, CT: Praeger Publishers, 2003.

Taito. *Space Invaders*. Taito (Arcade), 1978.

Team Ico. *Shadow of the Colossus*. Sony Computer Entertainment (PlayStation 2), 2005.

Team Meat. *Super Meat Boy*. Team Meat (Windows), 2010.

Teuber, Klaus. *The Settlers of Catan*. Kosmos (Board), 1995.

Thon, Jan-Noël. "Perspective in Contemporary Computer Games." In *Point of View, Perspective, and Focalization. Modeling Mediation in Narration*, ed. Peter Hühn, Wolf Schmid, and Jörg Schönert, 279–299. Berlin: De Gruyter, 2009.

Tolstoy, Leo. *Anna Karenina*. London: Penguin Books, 2003.

Ubisoft Montreal. *Far Cry 2*. Ubisoft (Xbox 360), 2008.

Valve. *Portal 2*. Valve (Windows), 2011.

Walton, Kendall L. *Mimesis as Make-Believe: On the Foundations of the Representational Arts*. Cambridge, MA: Harvard University Press, 1990.

Weiner, Bernard. "'Spontaneous' Causal Thinking." *Psychological Bulletin* 97 (1) (1985): 74–84.

Xbox.com. "Super Meat Boy." *Xbox Marketplace*, n.d. http://marketplace.xbox.com/en-US/Product/Super-Meat-Boy/66acd000-77fe-1000-9115-d80258410a5a.

Yanal, Robert J. *Paradoxes of Emotion and Fiction*. University Park: Pennsylvania State University Press, 1999.

Yates, James. "Carcassonne Strategy," 2011. http://www.chessandpoker.com/carcassonne-rules-and-strategy-guide.html.

Zichermann, Gabe, and Joselin Linder. *Game-Based Marketing: Inspire Customer Loyalty Through Rewards, Challenges, and Contests*. Hoboken, NJ: John Wiley & Sons, 2010.

Zillmann, Dolf. "Mood Management through Communication Choices." *American Behavioral Scientist* 31 (3) (January 1988): 327–340.

Zuckerman, Miron. "Attribution of Success and Failure Revisited, or: The Motivational Bias Is Alive and Well in Attribution Theory." *Journal of Personality* 47 (2) (1979): 245–287.

Zynga. *FarmVille*. Zynga (Flash), 2009.

Index